GW00455449

JAPANESE MOTORCYCLES

JAPANESE MOTORCYCLES

CYRIL AYTON

a Charles Herridge book

published by

Frederick Muller Limited
London

First published in Great Britain in 1981 by

Frederick Muller Limited
London NW2 6LB

© Copyright Charles Herridge Ltd. 1981

All rights reserved. No part of this
publication may be reproduced, stored in
a retrieval system, or transmitted, in any
form or by any means, electronic,
mechanical, photocopying, recording or
otherwise, without the prior consent of
Frederick Muller Limited.

Produced by Charles Herridge Ltd,
Tower House,
Abbotsham,
Devon
Designed by Bruce Aiken
Typeset by Toptown Printers Ltd., Barnstaple,
Devon.

ISBN 0 584 97075 7

Printed in Italy

Contents

The Postwar Scene and the Rise of the Industry

In 1945 there were fewer than 2,000 motorcycles registered in Japan. Production of two-wheelers, of any sort, that year amounted to no more than a few hundreds. Even pre-war, the Japanese makers had never managed more than 3,000 a year, when the total for Germany was about half a million. Yet by 1960 Honda alone were turning out more motorcycles than the combined manufacturers of Western Europe. J.W.E. Kelly has examined the Japanese motorcycle phenomenon in general terms.

Before World War Two (he explains) Japan was attempting to build a trading bloc in the Pacific, with herself at the head of it: a common market for the Far East. In going to war with China, she hoped to achieve this end partly by force of arms. An American embargo on the export of oil and other basic materials essential to the Japanese economy forced the country into world conflict, with the result that four years later, in August 1945, an American force landed on Japanese soil as vanguard of the occupation army. The early days were a revelation for both sides. The Japanese had fought with tigerish ferocity and the Americans expected resistance to continue, as guerrilla warfare, with equal intensity. Nothing of the sort occurred. The Japanese, for their part, were conditioned to expect Americans to exact revenge by pillage and rape. No such activities took place.

In the peaceful atmosphere of the occupation the relationship between victor and vanquished developed into mutual respect. Servicemen returned to the USA with a favourable impression, lowering any consumer resistance to the flood of Japanese imports in the 1960s and 1970s. In December 1945 Ambassador Edwin R. Rowley, head of the first reparations mission, said: 'Despite all the destruction, Japan retains in workable condition more plant and equipment than its rulers ever allowed to be used for civilian supply and consumption even in the peaceful years. That surplus must be taken out'.

But by January 1948 a reversal of this policy was clearly expressed in a speech by the Secretary of the Army in San Francisco: 'There has arisen an inevitable area of conflict between the original concept of broad demilitarization and the new purpose of building a self-supporting nation. Since last summer we have had a competent group of industrial engineers in the Pacific selecting the specific plants which . . . can be dismantled with the minimum of detriment to the Japanese economic recovery'.

The fundamental reason for the change of direction was the so-called Cold War in Europe. A Japan reduced to the standard of living of the early 1930s — the original intention of the Allies — would have made the country fair game for a Communist takeover.

Indeed, General MacArthur once referred to Japan as the Switzerland of the East. He saw Japan as a natural bastion against Communism. Not only were reparations kept to a minimum, but American aid began to pour into the country. Relief, mainly in the form of food, was supplied by GARIOA (government and relief in occupied areas). In 1949 the victory of the Chinese communists resulted in an escalation of the aid policy, and industrial raw materials were sent by EROA (economic rehabilitation of occupied areas).

American aid, allied to an excess of capital equipment and will to recover, formed a solid foundation stone to postwar economic development. Estimates published in June 1961 put the American aid figure at more than 1,700 million dollars, of which Japan had agreed to repay 490 million dollars over 15 years. In June 1950 the Korean war began, and gave another strong stimulus to the Japanese economy. An idea of the help the war gave to the country may be gained from the fact that in the period July to December 1950, contracts amounting to 184 million dollars were placed with Japanese industry. The war highlighted another factor which greatly assisted Japan's postwar recovery. Having been firmly demilitarized by the occupying powers, Japan was subsequently urged by MacArthur to send an armed force of 75,000 to help in Korea. The advice — order? — was resolutely refused, and in 1952 Japan stated its position in unequivocal terms: strictly defensive forces would be gradually built up within a rigid budget which the nation could easily afford. The lesson had been learned that less money spent on defence means more money available for industrial and social expansion. This policy was followed in later years, when the Vietnam war helped the economy without any question of Japanese involvment.

All these economic advantages enabled Japan to make a flying start on the task of satisfying the vast and almost untouched home market for personal transport. The trains, basis of Japanese commuter transport, were few and irregular, and buses and trams too were hopelessly inadequate. A bicycle appeared to offer the only means of getting about at above walking pace with some measure of independence. It was cheap and could be repaired indefinitely. But it was slow; and with too little food, perhaps even the patient Japanese found pedalling hard work. This was when people such as the resourceful Soichiro Honda saw their opportunity. He has told

Soichiro Honda, founder and president of the company photographed in the early 1960s.

Honda's Model A engine of 1948 ran on a fuel derived from pine tree roots.

how he hit on the idea of fitting an engine to a bicycle, 'simply because I did not want to use the incredibly crowded buses and trains and because it became impossible for me to drive my car because of the petrol shortage'. He began to fit the war-surplus 50cc Tohatsu engine into bicycles, and sold hundreds of them to cycle shops; and when supplies of the Tohatsu dried up he set about building his own engine.

Honda was only one — and one of the smallest — concerns providing cheap transport in the late 1940s. There were Rocket, Miyata, Sanyo, Ito and Showa, and any number of others, some building frames, others engines. For those selling complete machines it was at first, as for Honda, usually a matter of fitting a proprietary mini-engine into a bicycle frame. The most used engine was the ex-generator unit from Tohatsu, who were content to see some profit from the enterprise of all the tiny firms. Then, inevitably, Tohatsu woke up to the greater profits to be made by rehabilitating their own motorcycle line; thereafter their engines were earmarked for their own use. This marked a crisis point for the small makers, who had no option but to make their own engines or go out of business.

Honda's engine, a 50cc two-stroke, was built in 1948. Called the A model, demonstrating Mr Honda's orderly mind, it was a 1 hp two-stroke fixed above the front downtube of the bicycle and driving the rear wheel on a fuel derived mainly from the roots of pine trees. Not noticeably volatile, this brew would usually refuse to do anything significant until it had been churned through the engine for at least 15 minutes by Japanese leg power.

During this early postwar period Japan imported a trickle of foreign motorcycles, for copying purposes. The results were clear to see among the models produced by the 55 makers estimated to be operating in Japan around 1954/55.

At this time Japan was greatly helped by the *zaibutsu*, the trading companies. The *zaibutsu* were and are far more than trading companies in the Western sense. In Japan the majority of industrial firms apply themselves only to making the product; they leave sales, and often the buying of base materials, to a trading company. Over the years these concerns have developed an expertise in all matters relating to commerce, marketing, distribution, import and export, banking, insurance, transportation and research

into foreign markets. They have international standing and impeccable credit ratings, adding to the image of the product and facilitating raising capital. Often the trading company acts as intermediary with the bank and is able to re-lend to a smaller manufacturing firm unable to raise money direct. The cost of these services is generally very low, and the profits of a trading company are usually not in excess of ½ per cent. However, as the turnover is very large — Mitsui, handling Yamaha, had a daily turnover in 1973 of £18 million — profits are large, too, in absolute terms.

Government, banks and manufacturers, through the trading companies, work in consort in Japan to a greater extent than in the West. Ministries have a high reputation for efficiency; their planners maintain an analysis of the economic situation on a global basis, and thus are able to guide industry. Civil servants are sought for top jobs in industry or politics when they retire at the early age of 50. The Bank of Japan exercises a very positive degree of control. It is not an exaggeration to say that all the firms blend into one Japan Limited . . .

One result is a debt/equity ratio of about 80/20. Japanese management tend to fund a loan rather than comply with the wishes of shareholders, which enables the objective to be corporate growth rather than immediate profit. This philosophy pays dividends in establishing massive exports which, almost invariably, are seen within a year or two as no better than 'dumping' by those countries chosen to receive the Japanese onslaught and thereafter finding their own industries in trouble. On dumping, Dr Kelly makes the point that an economist defines the term as selling at below prime cost. But with a vast home market, yielding economies of scale beyond the wildest dreams of (for example) British manufacturers, first in motorcycles, later in cars, marginal costing may be used with impunity. Tooling and other costs of a fixed nature can be absorbed by a spread over the home market, enabling the maker to export at slightly over prime cost and still show a profit. It may be called a sacrifice, but it has been shown to be an effective way of killing competition. When the rivals have disappeared, prices can be raised.

Many of the bigger Japanese firms in the late 1940s were, like Tohatsu, merely picking up the threads of their pre-war activities. Cabton, for instance, founded in 1934 and the most successful Japanese manufacturer up to 1940, had long specialized in big singles. A few years after the war they climbed to third place in the motorcycle industry with machines of never less than 250cc, and usually considerably larger: 350, 500 and 600 singles and twins. By 1953 their most widely sold model was the RTS 600 vertical twin, which gave 28 bhp, had a four-speed gearbox and was good for 80 mph.

Tohatsu was a more typical manufacturer. Returning to motorcycle production in 1948, the company hoisted itself to the position of market leader. The most popular Tohatsu was the 60cc 2 bhp Puppy, an enclosed two-stroke with two-speed transmission. An indication of Tohatsu's strength in the postwar scene is provided by the 1955 production run of 72,000 units, when Honda were turning out no more than 32,000. Tohatsu introduced a rotary-valve single in 1958 and a 6.7 bhp 50, the Runpet, which could reach 75 mph. Exports to the USA and Europe were brisk, but after some years Tohatsu designs were showing a staleness which began to affect sales. Yamaha and Suzuki moved ahead in 1960, when Tohatsu production dropped to 40,000 units. But the firm remained in being into the early 1960s until a year-long industrial dispute brought production, and exports, to a halt and forced it out of the motorcycle business. In 1963 a Tohatsu was the chosen machine in the UK of Richard Wyler, an American TV actor of modest talent. Bad luck (or simple justice) had denied him top rewards in his profession. Possibly for consolation, he turned to motorcycle racing, and appeared at Brands Hatch and other venues with his beautifully prepared two-stroke. It was a two-carburettor 125 twin with expansion-chamber exhaust system and 12-volt coil ignition. Maximum power was undisclosed but arrived at 11,000 rpm, and Richard Wyler claimed a top speed of 110 mph.

Another big name from pre-war days was Rukuo, founded 1935, who specialized in Harley-Davidson lookalikes (legally, under licence) until Pearl Harbor and thereafter carried on in exactly the same way but without putting a percentage aside for the American company. After the war they continued, with no perceptible updating of the H-D design, but cushioned by contracts to supply the sv twins to police forces, until senility exacerbated by competition from Meguro, a livelier firm, overtook them in 1960.

The Meguro 170cc Ranger, a neat ohv single which sold well in Japan in the 1950s. Meguro started in business in 1937 and were taken over by Kawasaki in 1961.

ebb. After the war, Meguro remained faithful to the big single, later adding 500 and 650cc twins bearing a startling resemblance to the BSA A7 series. In the early 1960s, with sales dropping, Meguro were taken over by Kawasaki.

Bridgestone was part of the giant tyre-making concern, and extremely successful in the 1950s with two-stroke engines produced for other manufacturers to fit in their frames. By the end of the decade they were turning out complete motor-cycles of high quality, showing considerable innovative thinking, with rotary valves and disc braking long before a disc appeared on the Honda CB750. But Yamaha and Suzuki had been single-minded in their attack on the two-stroke market and were able to downgrade the Bridgestone until, in the early '70s, the company retired from the contest.

Almost a decade ago I.A. Thickett, close observer of the mushrooming Japanese motorcycle industry, singled out Honda for special attention. What determined him to sort myth from reality was a television interview in which captains of British industry had dis-missed the threat posed by the Japanese. Dennis Poore, chairman of Norton Villiers, claimed that utility machines had almost ceased to exist in the UK. Eric Turner, accountant turned BSA chair-man, who regulated his life with a fresh carnation daily and new Rolls each year, declared that the Japanese copied British big bikes. Thickett pointed out that in only a couple of cases did Japanese motorcycles over 500cc match a corres-ponding-size British bike, even to the number of cylinders. And dealing with Mr Poore's contention . . . obviously the Norton man had not thought to take his 6.3 Mercedes through any busy city, to view the ever-growing number of Honda 50s, and 90s, and 125s. Honda's success came down to marketing, according to Thickett.

There were other firms, such as Subaru and Mitsubishi, later to grow to great size and standing outside the motorcycle field, often as car makers. Subaru began making motorcycles after a company engineer had tested a Powell scooter used by US paratroopers. The Subaru Rabbit, fashioned after the Powell, remained in production for 22 years, until 1968, when the firm decided to drop two-wheelers in favour of all-out car production. Mitsubishi's two-wheeler was the Silver Pigeon, which was copied from the American Salisbury scooter and lasted until 1962, when full attention was turned to the Colt car.

Meguro, established in 1924, was a founder member of the Japanese motor-cycle industry. The firm enjoyed a high reputation with a 500 single which was an obvious derivative of the pre-war Velocette MSS and, like the Velo, that prime example of conservative design and engineering, was well made, hand-built almost, and happy to run for many years with very little attention — which was fine for the paying customer but kept the company's financial resources at a low

The word marketing is bandied about as a magic formula. People tend to think that setting up a marketing service or market research department to turn out endless streams of statistics is all that is needed to guarantee a firm's profitable existence. Of course this is not true. Successful marketing is practical; it doesn't stay in the realms of theory and speculation; it looks everywhere, at the market and at the company and at the com-petition. It observes, records and makes decisions on the basis that as many relevant factors as possible have been considered; whether information has been supplied by the market research statistician or the works manager is of no importance.

Honda

Soichiro Honda was fortunate in teaming up with Takeo Fujisawa, a super-salesman who was to make 'marketing' his side of the business. It was 1949, and Honda had been operating for three years. He was selling about 100 units — light motorcycles and mopeds — a month. The market was vast. But how to tap it? Honda was an engineer and something of a designer and, on contemporary evidence, no more than a small-scale manufacturer. He was not a commercial genius. Then along came Fujisawa, who was at least usefully near to being one. Honda recognized in Fujisawa a quality which he, Honda, did not possess. Being a sensible man, he put Fujisawa in charge of sales, which meant that almost from the beginning the Honda company was sales-orientated, the management function in the hands of a salesman who could put his point of view forcibly. By 1953 Honda was the leading motorcycle producer in Japan. But success in Japan was seen as a stepping stone to exporting to the world. Honda went to the USA and to Europe in the mid-1950s, and realized

then that his products did not offer effective competition to foreign machines founded on generations of high-quality workmanship: the Hondas, by contrast, were crude and old fashioned. The Japanese government demonstrated long sight by providing low-interest loans for the purchase of new machine tools from America and Germany. With the co-operation of his workers, who subsisted on low wages and cut their holidays, Honda produced new motorcycles at a furious rate and soon repaid the loan.

But exports were a tough nut. In 1951 the Japanese motorcycle industry (all 80 firms) had produced 24,309 machines and in 1954 164,477. Attempts were made to sell abroad but failed miserably, sales declining from 1,113 in 1948 to 318 in 1951 and 141 in 1954. Even the most perceptive, or fearful, of European manufacturers appeared to have little reason to trouble himself with visions of Japanese imports pouring into his country. The Japanese, it seemed, would be busy for generations getting their own people on to the roads. And that, of course, was a grave error of judgement for the Europeans, an early sign of the complacency that was to inhibit their responses to Japanese competition in later years.

Rare Honda — the 350 four. Slightly enlarged, it became the CB400 and won a loyal following in Europe.

The Honda Cub step-through first appeared in 1958. This handy, simple runabout put much of the world on two wheels and made the terms 'motorcycle' and 'Honda' apparently interchangeable. By 1967 no fewer than five million Cubs, in one form or another, had been sold. This is the C240 Port-Cub of 1962.

August 1958 may be seen as the start of Honda's rise to world leadership, with the introduction of the step-through 50cc Super Cub which continues, in updated form, to be the staple of the Honda motorcycle programme. Exports began in 1957 with five machines and 12 years later were only a few thousand under a million. In 1969 40 per cent of powered two-wheelers sold in the UK were manufactured by Honda; in 1980 46 per cent. In the USA and Japan Honda's share has been even higher.

above
A 10,000 rpm, 95 mph two-fifty in 1960: the ohc CB72 Honda Dream.

right
Honda's early ideas of a cross-country motorcycle ran to little more than altering a CB72 by fitting a smaller tank, high-level exhaust pipe, steering damper and dust-excluders for the front fork. A good road bike: hopeless on the rough.

16

With Honda leading the other Japanese makers in the export drive, serious thought was given to the 'carrot' that was to persuade the American motorcyclist to forsake his dearly loved Harley (and Norton and BSA), the German his beautiful BMW, the Englishman his fast, vibratory Norton, the Italian his Moto Guzzi. Only minimal research was required to show that all were influenced by well publicized sporting successes. The man with a humble 125 BSA Bantam, getting to grips with the full horror of petroil lubrication, 'whiskering' plugs and derisory hill-climb performance, would know all about the latest BSA Gold Star successes in the Isle of Man Clubman's races. In the States a would-be Brando left gasping at the lights by a family Chevvy might dream that his Harley 'flat head' had lapped Daytona at one-o-one. But road-racing was very expensive. English factories had found the cost too high, as had the Germans and Italians — with the exception of Count Agusta, who continued to derive perennial satisfaction from watching John Surtees, and later Mike Hailwood, circling the race circuits of the world in lonely splendour on the MV four. Not that sales of the Count's bread-and-butter line in motorbikes appeared to profit from his racing wins. But he was a millionaire with eccentric tastes, and motorbikes were only part of his industrial offerings; and perhaps there had to be an exception to prove this new-found Japanese rule, that racing sold motorbikes. Racing across country — moto-cross — sold motorcycles; and so did the feet-up game of trials riding. The Japanese were to get round to these competitions a little later.

top
The first Honda four to be brought into the UK, in 1968, on view to journalists at the Charing Cross hotel. The CB750 ushered in the era of the so-called superbike.

above
Offside of the CB750. Power of the sohc engine was 67 bhp, top speed of the 480lb bike nearly 120 mph.

left
The CB350 was externally identical to the CB250, which was a 1969 successor to the CB72. A dependable, unexciting machine, the 350 was Honda's biggest seller in the USA but fared less well in Europe.

17

For the time being it was clear that the greatest impact on the international scene was to be made through road racing. It would be expensive, but Mr Honda was prepared to spend a great deal of money; at the end of the 1950s his share of total Japanese motorcycle production had risen to 40 per cent and he had made a great deal of money . . . By way of victory in the Catalina races in the USA, the Honda team journeyed to the Isle of Man for the 1959 TT races, and thence to years of almost complete domination on the circuits of the world, as recounted in a later chapter.

Honda's thinking about the race effort, its importance in boosting the sales of everyday machines, is shown in one of his pep talks to the work force: ' . . . we have now definitely established ourselves as the world's top-ranking motorcycle manufacturer. The explanation of this remarkable development and expansion of our firm lies to a great extent in the complete understanding by our 5,500 employees of our policy: namely, to manufacture outstanding products capable of superior performance from a world-wide marketing point of view, for sale at the lowest possible price, and made to meet the requirements of customers elsewhere . . . a universal principle which should be considered and put into practice by everybody in every place irrespective of nationality'.

While Mr Honda was busily engaged in making his million units, and exporting 60,000 of them, the British industry came up with approximately 90,000 in total. These of course would be what BSA executives called 'real' motorcycles and, when carried away by strong British ale and deep-dyed xenophobia, 'real men's' motorcycles. H.M. Palin, then director of the British Cycle and Motor Cycle Industries Association, said that Britain exported in 1962 3½ times as many two-wheelers of all types as her nearest competitor and four times as many as Japan.

above
Honda's 400-4 of the late 1970s was very popular in the UK but 'died' in America — then, as ever, the main market for the Japanese and thus arbiter of manufacturing trends.

left
The Gold Wing, introduced in 1974, with water-cooled four-cylinder engine and rear shaft drive, was long, heavy and extremely quiet both mechanically and on the exhaust. It remains so today, with an increase in cubic capacity and improved handling. The 'tank' is opened to show the electrics; fuel is carried under the seat; all in the interests of a low centre of gravity.

opposite
Honda's 297cc six

19

above
Almost 10 years after the
CB450 was dropped Honda
brought back a dohc parallel
twin, the CB500T, which
appeared in the USA in 1974.
It vibrated too much, was not
particularly fast, suffered by
comparison with the
medium-size fours. Sales
even at discount prices were
low and today the 500T is
rememberd more for its
'colour coded' brown dualseat
than for any technical merit.

right
The Super Dream Honda,
Britain's most popular 250.
This is a 1980 N-A model.

What Mr Palin — who returned in the 1980s to something akin to his 1962 job after years in the industrial wilderness fighting the Orientals — did not do in his inspiriting message was particularize the British effort by category or even by type of power, mechanical or human. The giveaway was his definition, 'two wheelers of all types', which almost certainly covered pedal cycles of which Raleigh, for one, made a great many. If

Honda's exports of 125cc-and-over motor-cycles amounted to no more than a quarter of total exports, 15,000 among 60,000, which would be in line with the total production breakdown, then indeed Britain was ahead, for in 1961 BSA and Triumph, Norton and Royal Enfield and the rest sent abroad approximately 29,000 machines. But by the following year that level would have declined, possibly by up to 5,000, and the Japanese effort would have strengthened.

Dealing with this welter of statistics a motorcycling expert from Holland, Ger-hard Klomps, pointed out that though pessimists in Europe who cited Honda's million-a-year production might be taken aback by the sensible breakdown into 'real' and motorcycles and mopeds, there could be little comfort for the West even in a revised figure which set Honda's output of 125cc-and-above machines at no fewer than 20,000 a month, a quarter-million annually. Klomps wrote: 'It is en-couraging to learn that England still has greater exports in two-wheelers than Japan . . . her lead is something that could be overlooked under the bombardment of Japanese statistics . . .' Unfortunately for the 15,000 at BSA Small Heath and the other thousands employed in European motorcycle-making, their leaders obvi-ously found Mr Palin's words too en-couraging, and it is history that very little effort was made to match the Japanese.

above
The 1978 dohc Hondas, of 900 and 750cc, were introduced to Europe in West Germany.

left
A new generation of high-performance Hondas appeared in 1978: dohc 16-valve fours with 140 mph potential.

Mr. Fujisawa, talking to a gathering of dealers at the Honda plant at Suzuka, said: 'When you make a comparison between the US and Japan, based on the volume of sales, you will find much in common — today's US market is about the same as Japan's eight years ago. I assume that you have observed that Honda's progress is very much like that of Ford in the early days. America had nobody in the two-wheeler field who could be compared to Mr Ford. It is quite obvious why the two-wheeler market in the US made no advance in the past. Honda's exports to the US in 1959 were 839 units, and 5,082 in 1960. This year we expect to send 80,000. In Japan the registration percentage was one machine per 900 people in 1950, and now it is one for 20. I doubt if anyone in the US predicted the popularity the motorcar would gain one day. Therefore you cannot say that it is impossible to bring the sales volume to the same level in the US as in Japan. One in 20 for the USA means around one million Americans owning motorcyles — Japanese motor-cycles. That figure was passed midway through the 1960s'.

Fujisawa had wanted to get into the American market from his earliest days at Honda. The USA, it seemed, had put the motorcycle behind it; they were play-things — and, being predominantly 1,200cc 600 lb Harley-Davidsons, not very handy playthings. What few imported machines there were came from Britain. They were more up to date than the Harleys, having ohv engines and more nimble handling, but 'sophistic-ated' would not be an appropriate word to use in describing them. They leaked oil, were noisy and had to be kickstarted. Fujisawa thought the USA might turn out to be a pushover for Honda.

American Honda began business in Los Angeles in 1959 with an investment of a half-million dollars. Where it would have been natural for Honda to send out Japanese personnel, the decision was taken to employ Americans, in the way that a couple of years later a British outpost for Honda at Kingston on Thames would take on British personnel. Fuji-sawa and Honda realized that less res-entment would be aroused by conducting their business abroad like this than by flying in teams of alien, hard-working Japanese. Apart from any question of commercial friction, World War Two was still comparatively recent history . . . Suzuki were to take this marketing philo-sophy a stage further in the UK by engaging Associated Motor Cycles, one

The CB250 Honda has been the best selling 250 in the UK for some years. Rather overweight, and large (it is indistinguishable from its larger-capacity stablemate, the 400), this sohc twin reflects Honda's acute analysis of what 'Mr Average' looks for in a motorcycle.

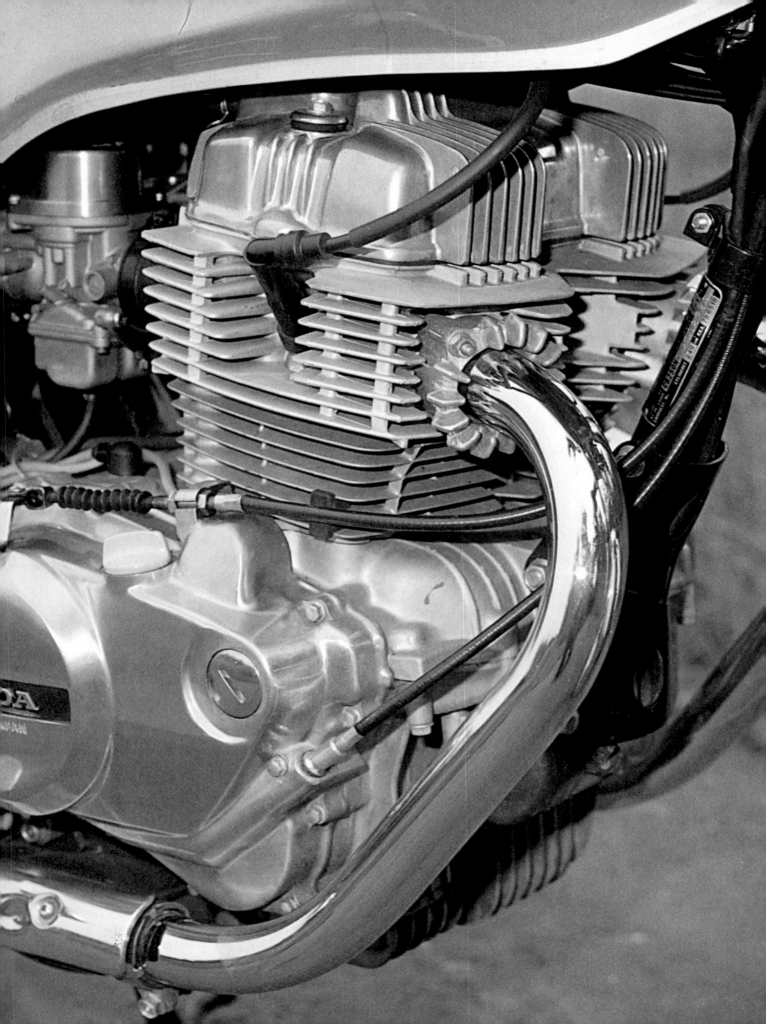

of the largest of indigenous manufacturers, to sell their two-strokes on the reasonable grounds that any industry wrath at the newcomers would be dissipated by the home side's share in profits. That Suzuki did not carry out their homework sufficiently well to reveal the shaky state of AMC's affairs, which shortly led to the group's collapse and Suzuki's imports coming to a halt, does not invalidate the good sense of the plan.

Honda in America displayed drive and innovative thinking in promoting the C50 step-through. Where before motorcycles had been advertised only in specialist papers, the step-through was featured in enticing full colour among familiar advertisements for Magnavox and Coca Cola in national magazines such as *Life*, *Time* and *Look*. Honda's agency was Grey's Advertising of Los Angeles, who deserve much of the credit for this new approach. The advertisements showed Hondas being used in any

number of ordinary domestic situations by clean-cut businessmen who were often of middle years, and only a shade out of the ordinary by virtue of an income big enough to support a Honda in addition to the mandatory gas-guzzling automobile. Even American housewives rode Hondas, or would do so given half a chance, suggested Grey's.

Everything was calculated to upset the idea that motorcycles were exclusively for anti-social misfits who wore leather jackets and spoke in early Brando. The slogan 'You meet the nicest people on a Honda' was born around 1960 and was to endure into the 1970s. Within three years Honda had a nationwide network of dealers in America. By 1968 a million Hondas had been sold and the company was close to taking 50 per cent of the total motorcycle market. 'Honda' passed into American idiom. Films had characters played by stars like Jack Lemmon saying 'I'm buying a Honda'.

Honda 750 (early 1970s version) at Showtime.

Frame of the new fours was well braced and a notable improvement on earlier Honda efforts. The result was good handling, even with the 90 bhp 900cc engine installed, at well over 100 mph.

Honda six. The experts prophesied 150 mph top speed when the bike was first announced, in 1978. They were 20 mph to the good.

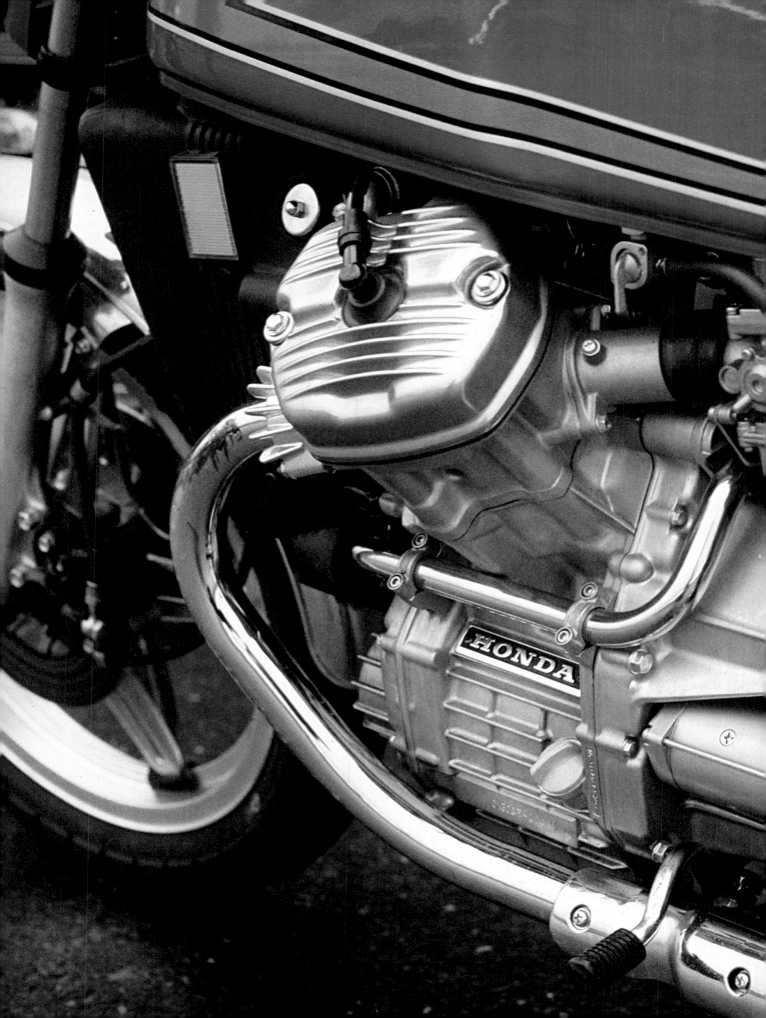

Honda were told that motorcycles sold in the USA to young people — possibly an unsurprising item to be thrown up by research. This, however, is what advertising agencies are employed to do ... to substantiate, at considerable expense to the client, the client's own reasonable assumptions about his product; at which point, bolstered by irrefutable facts and figures, he may feel entirely justified in putting in train plans to spend a great deal of shareholders' money. Having established that step-throughs were suitable for the large part of the US population which was middle-aged or female, or both, Honda went all out to win over the 'respectable' — i.e. non-Hell's Angels — young who might want to take a motorcycle into the desert or the canyon or the forest; this sort of usage effectively ruling out most of the eastern seaboard and serving to confirm that motorcycles are best enjoyed in a benign climate like that of California.

Another best-seller from Honda: the CX500 water-cooled vee-twin with rear shaft drive. Again, like so many Hondas, this one is a little ponderous with, inevitably, a fair amount of high-up weight.

above
Still to come as a production motorcycle, Honda's turbo-charged CX500 was shown in prototype form at the 1980 Cologne Show.

Suzuki

Suzuki was an old (1909) established company that for many years engaged solely in the manufacture of weaving looms (though there was, in the late 1930s, some experiment with a small car

right
Jitsujiro Suzuki, President of Suzuki, in 1973.

based on the Austin 7.) After the War, with large, if damaged, factories and a work force running into thousands to keep busy, the head of the firm, Michio Suzuki, turned to two-wheelers as a way of utilizing labour and turning his American financial backing to good account. Unlike Mr Honda in his time, Suzuki determined to produce his machine for the masses in its entirety; there was to be no buying-in of engines.

The first Suzuki engine, a 36cc two-stroke, was installed in a bicycle in 1952 and was sold as the Power Free, a name which Suzuki historians profess to find entirely logical because at the time it implied (I think) that the power to get along at 30 mph was free of human agency; or something like that. The engine was mounted in the middle of the main frame and drove the rear wheel by chain through a double-sprocket arrangement which allowed the rider to weigh in with light pedal assistance as he saw fit or terrain demanded; or to take over entirely, cutting out the engine; or of course to leave everything to those hard-pressed 36ccs. Two-wheeler business at Suzuki went ahead in good style. There

* Etymologists with an interest in the Japanese tongue may care to ponder later developments of Suzuki's fixation with '-Free' in the names of those early bikes. What about Mini Free (b.1954), a moped? Or Porter Free (b.1955), a 100cc-powered sidecar outfit?

Suzuki T20 Super Six in 1966. This one brought two-stroke performance within a few mph of the standard set by the ohc CB72 and was regarded as having the best all-round handling of all the 250s. Power output was 29 bhp at 7,500 rpm — high enough to make the T20 a useful competitor in long-distance racing. In the USA the T20 was known as X-6.

above
After the 500 twin, Suzuki became hooked on the idea of three cylinders — water-cooled or, as here, 'ram air' cooled, which meant having a lid across the separate cylinder heads to direct air where most needed. This is the 37 bhp 380cc version; another was a 550; both first appeared in 1972.

The Japanese manufacturers have their own test tracks. This is Suzuki's Ryuyo 4.04-mile circuit by the sea.

were cash subsidies from interested parties such as the Japan Patent Agency to boost research into improvements to the Power Free, with the result that a 60cc two-speed Diamond Free ... I have yet to hear of a remotely convincing explanation for this title ... made an appearance, to be involved in publicity stunts which helped to lift total production to a reputed (probably untrue) 6,000 units a month.

This was 1953, when the motorcycle side of the business was not known as Suzuki but as SJK (for Suzuki Jidosha Kogyo). It was not until 1954 that Mr Suzuki decided he had had enough of supplying 'Free' motors to the small-time frame makers who traded on Suzuki workmanship and collected most of the credit, and the customers' money. The first all-Suzuki two-wheeler came out in May 1954 and was called Colleda, an attractive name which translates as 'This is it!' — with some dispute among experts as to the appropriateness of the exclamation mark. Certainly 'This is it' pronounced on a falling or, worse, interrogative note probably would not reflect Mr Suzuki's intentions when 'Colleda' was chosen. Apart from being the first all-Suzuki (even if it was an SJK) motorcycle, the Colleda is interesting in the history of the company because of its 90cc engine, which was a 48 x 50mm single-cylinder four-stroke. In those early days Suzuki were still in doubt as to which path to follow: the two-stroke was crude and unreliable, the four-stroke expensive for a cost-conscious manufacturer. The Colleda CO was, as it turned out, the last four-stroke to be built by Suzuki for 20 years. It had a three-speed gearbox in unit with the engine; an oil-filter; and automatic advance for the ignition. The front fork was telescopic and the pressed-steel frame had a plunger system of springing.

By 1955 Suzuki were back to two-strokes with the Colleda ST, which had a 52 x 58mm single-cylinder engine with cast-iron barrel and alloy head. It was sold at first with a rigid frame, on account of unflattering comments about the low-comfort, high-wear characteristics of the plunger arrangement fitted to the first Colleda. This two-stroke was a success and was developed over the years to the end of the decade, acquiring pivoted-fork rear springing and numerous other embellishments to keep it competitive. In 1956 the first two-stroke twin was announced. The Colleda TT was a 250 with pressed-steel frame, pivoted-fork springing, Earles-type front fork, a four-speed gearbox and 7in-diameter brakes

in full-width hubs. It was nicely made, smoothly styled; and it was made even more attractive the following year when the leading-fork steering/suspension was changed to telescopics and the pressed-steel frame replaced by duplex tubes.

If the early Colledas were tough, no-nonsense workhorses, rather like East Europe's CZ, the twin — it even had winkers — had much of the svelte appeal of the best of West Germany's machines such as Adler and Victoria. The twin produced about 16 bhp, which was enough to give a top speed of 80 mph at a time when an old-fashioned device like the 1,200cc Rukuo-made Harley-Davidson was labouring at little more than 70 mph.

The good reception for the new Colleda, followed by gratifying sales, encouraged Suzuki to introduce automated assembly lines into their new factory, on the pattern laid down a year or

Suzuki made up for the Wankel engine fiasco by bringing out a range of extremely serviceable, if conventional, four-stroke multis with dohc valve operation. The 750-4 was marketed first and was followed by the 1000 which set high standards of handling in a class traditionally strong on power but weak on stability.

In the early 1970s there was brave talk from Suzuki about their revolutionary rotary engine. But the RE-5 proved to be an expensive flop. After a couple of years of mediocre sales it was withdrawn from the market. Racing excursions were, despite the evidence of this picture, extremely rare and not very successful.

above
One of Suzuki's innumerable runabouts of modest size, and performance, that appeared through the 1970s. This is the B120 of 1971.

right
Suzuki's R and D: up-to-the-minute equipment and unlimited cash . . .

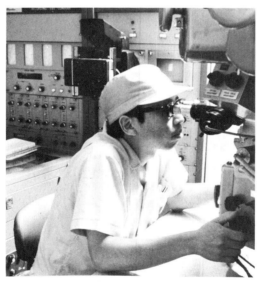

so earlier by ambitious Honda. Around 1960 the Colleda lost the SJK trade mark and turned over to the characteristic 'S' for Suzuki Motor, and was re-styled, with a change to a back-bone chassis (the engine depending from it), valanced, stayless front mudguard, and horseshoe-shape headlamp. This last is a styling fad that has figured on assorted Suzuki models up to the present time, thrilling generations of stylists in Hamamatsu, all of whom have been infatuated with its 'different' shape. Some have said that a Suzuki's headlamp beam at night can be distinguished from all others — a comment which might surprise anybody who actually ventures on to the road at night on a Suzuki, and infuriate any other body who has been stranded with a broken 'horseshoe' and thrown on to the mercies of a stockist of conventional headlamps.

By 1964 Suzuki were second to Honda on the home market, with 17 per cent of total sales, and were leading the world as two-stroke producers. In 1961 the company had opened a London office, and within a couple of years were sending into Britain 17,000 machines annually. In the USA Suzuki set up a marketing agency in Santa Fé, California, and by the end of 1963 were the second motorcycle importer, behind Honda. Other outlets were created in Singapore, Taiwan and Pakistan. By way of outstanding two-strokes such as the 29 bhp, T20 (XT-6 for America) 250 twin of 1964, with its 90-plus mph performance, six-speed gearbox and separated oiling, the 1967 500 twin, and the later threes, both water- and air-cooled, Suzuki progressed towards a resumption of four-stroke production, in 1975/76. With a minor hiccough on the way...

In the 1980s Suzuki executives tend not to hear, at first, any question envolving the RE-5 rotary. They believe that anybody brash or naive enough to bring up the subject merits a little of the 'I

RE-5 at Hamamatsu. The rotary came out in a blaze of publicity, stuttered along at moderate speed and high mpg, and disappeared very, very quietly.

Suzuki's first big-production four-stroke, the dohc 750-4 of 1977. Four years on, the GS750 remains recognizably the same motorcycle apart from an extra disc.

The X-7 250 two-stroke twin — brought in as a final fling in the two-stroke effort by Suzuki in 1978. Light and for a 250 very fast, with a top speed tantalizingly close to 100 mph, it was aimed squarely at the market enjoyed by Yamaha's long-established RD series. But missed, apparently.

don't understand/hear' treatment. If as would-be interrogator (I offer advice because I have attempted the role) you stick to your questioning, you will have an answer, finally. It will be brief. Obviously there has been agreement on the response most likely to succeed in killing further interest, for a standard one is used in various countries, by various people. 'The RE-5', they say, 'was a considered exercise to demonstrate Suzuki mastery of engine technology. *And would you care for more sake?'*

Nobody in engineering cares to admit to boo-boos. I have heard that the Edsel motor car was seldom mentioned at dinner parties at the Henry Fords in the 1950s. And Americans, of all people, are ready to analyse their own mistakes. The Japanese have no such tendency. Mistakes do not occur. Or if they do occur (Hamamatsu presumably being only a shade less fallible than Detroit), they can be reclassified as 'exercises to demonstrate . . .'

I rode a prototype RE-5 in Japan in 1973. Suzuki then were very proud of it. They let me know I was privileged in being allowed to have a ride. When it was marketed in 1974, the RE-5 represented a peak in a policy summed up in the widely used, if oddly phrased motto *Make Only Valuable Products*. Fruit of the Suzuki Technical Centre's '1,000 productive minds and the latest computer system', the new engine employed a single rotor, was described as 497cc, giving 62 bhp, and had a fuel consumption averaging 30 miles to the gallon of 2-star. It was rather heavy and not very fast by ordinary 750 standards. What killed it, however, was the despair that engulfed most of the likely owners, and even the unimpressionable dealers who were supposed to sell the bike, when they considered the pitfalls and the costs involved in maintaining, never mind repairing, the monster. Sales in Britain, where Suzuki had great hopes of the rotary, amounted to no more than 250 in the three years to 1977, when it was withdrawn from the market and from promotional literature.

A little later that year Suzuki assuaged their pride with a new line in old-fashioned reciprocating-piston technology, this time on the four-stroke cycle. It was clear that they had suffered in some part of the corporate soul, for these four-strokes were as determinedly 'mainstream' as the RE-5, and some of the earlier two-strokes, had been innovative or at least halfway original. The GS750 four was very 'CB750', heightened by a dash of Kawasaki Z1 thinking by way of

THE NEW GSX 250

double overhead camshafts instead of single. The GS400 was part of the package and, predictably, had little of the big four's excellent performance, or appeal. From this two-model base, the company went on to manufacture a series of across-the-frame twins and fours, all dohc, the latest having four valves per cylinder, and some with shaft drive, that have enjoyed a deservedly high reputation for above-average handling and performance.

Asked in 1979 if there was any intention to build a larger bike than the 997cc GS fours then near the top of the 'super-bike' class, an executive in Suzuki's overseas operations division replied, 'If Suzuki was to decide to make a big sports tourer with shaft drive, then we might have to bring up displacement a bit to make up for the power loss driving the shaft'. Four months later the 135 mph GS 1100 ('sports tourer' is hardly the right description for it), with chain drive, made an appearance in Europe; it had — of course — four-valve heads ('. . . there are great possibilities for performance latent in the two-valve head design'); and in short a cynic might be forgiven for seeing Hashimoto, Yokouchi and the rest* as company men promulgating an impenetrable line of company propaganda only occasionally related to fact or intention.

More in keeping with these emission-conscious times was the GSX250, a four-stroke twin with four valves per cylinder. Definitely a more sporting proposition than the Honda Dream, this fast Suzuki has yet to disturb the market leader.

opposite
Natural progression for a manufacturer selling a popular chain-drive multi appears to be to go to shaft drive, with an increase in engine capacity by at least 100cc (to compensate for the shaft's power absorption) and a jump in selling price. This is the GS850, development of the original 750-4.

* As interviewed by C.D. Bohon for American *Motorcyclist.*

Yamaha

Around 1960 Honda were producing 52 per cent of Japan's motorcycles, 930,000 units among 1,800,000. There were still, apart from the clear leader, no fewer than 16 makers to make up the 100 per cent, some as small as 0.1 per cent of the market, and probably countless tiny firms below that point. While much of Honda's total was made up of 50-125cc machines, as detailed earlier, Yamaha with their mainly 125cc-and-over bikes were up at number four, on 12.9 per cent, well ahead of Kawasaki/Meguro, 12th in the list. It is appropriate, therefore, to consider Yamaha, which began as Nippon Gakki late in the last century and carried on happily enough as Japan's sole musical instrument manfacturer until 1950, when Genichi Kawakami took over. Five years later, with a fine eye for business, Kawakami announced the formation of a Nippon Gakki offshoot that was to produce motorcycles under the name of Yamaha Motor.

The first Yamaha was a 125, named YA-1. There is uncertainty still as to whether Yamaha copied the British BSA Bantam two-stroke, which was a seven-year-old copy of the pre-war German DKW, or went direct to the original. DKW and BSA were highly efficient two-strokes and sold well, and the same may be said of the YA-1, which went on to win a race at Mount Asama, Japan's premier race meeting, on its first outing. The YA-1 had a maroon and white finish and clean, light lines, with a telescopic front fork and plunger rear springing. The engine had 52 x 58mm bore and stroke and was claimed to put out 5.6 bhp at 5,000 rpm, which translated, courtesy of a four-speed gearbox, into a top speed of 50 mph, in line with DKW/BSA performance. Next bike off the Yamaha line was a 250 twin, the YD-1. Again Yamaha displayed sound business sense (and possibly a disregard for business 'ethics') by copying another German design, the two-stroke Adler. When it came to England in the mid-50s, the 250 Adler had shown that it was at least a match for so-called sporting 500s, designed 20 years before, and far too good for the home-produced 250s. There was no sign, however, that the British makers were concerned by German superiority. Home and export sales were reassuringly buoyant, and British riders at that early date were still locked in the xenophobia that was only to melt away with the coming of the Japanese. There was no

Yamaha's YD3 250 twin of 1958 was a smoother version of the YD1 that was the company's fist significant success. The smooth, flowing lines — the slightly 'bland' look — were to be modified, and the enclosure for the rear chain dispensed with, when Yamaha realized that motorcycles of around 250cc appealed to people with sporting ideas.

left

The YDS series of Yamaha 250s gave 90mph-plus performance in the mid-1960s. Engine power varied from 25 bhp for the YDS-3 of 1965 to 29.5 for the later YDS-5. Electric starting and Autolube oil injection were de-luxe touches. Altogether, there was something of the racer about these well-made, sporting two-strokes. This is the YDS-3.

below

A YDS-7 showing the development of Yamaha's 250 line, with angularity replacing the curves of the earlier models. The YDS-7 of 1972 became the RD250.

Yamaha, like Suzuki, began to ease two-strokes out of its range in the late 1970s. The 1100 four made the point, very firmly, that Yamaha could match anybody in four-stroke technology. It was introduced in West Africa.

question of other makers 'doing a Bantam', which might have been sensible. That uncharacteristic plagiarism had been more a political act, under the blanket of war reparations, than the result of any commercial acumen. The DY-1 was taken to Mt Asama and there repeated the 125's winning record. Production soared, and so did profits. Other manufacturers were overtaken in the race behind Honda, unassailable in the lead, and at least two of them were swallowed by Yamaha. One was Showa, once number two among Japanese makers, who had produced advanced two-strokes with reed-valve porting, one with a mono-coque body, Earles-type front fork, total enclosure for the chain drive, tubeless tyres and push-button gear changing.

The other firm taken over by Yamaha was Hosk, which had as mainstay of its limited range an overhead-camshaft 500 twin. This bike had a top speed of 110 mph and was like not one but several contemporary British twins (apart from

its ohc operation, of course, which stood in British circles for High Technology and was, accordingly, reserved for the special demands of racing). Hosk took the Japanese talent — near genius — for copying to an interesting stage by creating a composite from all that was best in the British industry. It was very successful on road and track. Thus Yamaha acquired considerable extra expertise, and the result was seen when they brought out a four-stroke, under the Yamaha name, in 1969. The ohc 650 traced its antecedents through Hosk's 500 to any number of British parallel twins, not least in its outstanding characteristic of hand-numbing vibration that was as severe as anything suffered by owners of early 1950s BSAs.

By 1959/60 Yamaha had refined their two-strokes, turning out a 250 twin, the YDS, with five-speed gearbox, which produced 18 bhp, had a true top speed of 85 mph and embarrassed a number of Honda fours at the 1959 Asama races

before cracking up. This 250 was the basis for the racing twins of the 1960s, the RD56s, first air-cooled, then with water jackets, which were always to look absurdly simple — 'standard' — when contrasted with the multi-cylinder complexity of the Hondas. Yamaha's standing as a two-stroke manufacturer, of road machines as well as racers, derives from the YDS. The grand prix vee-fours of the late 1960s, brought into the arena to match the fives and sixes (and a rumoured eight) from Honda, were in a sense aberrations — weird and wonderful artefacts that sprang from the drawing board, without background or much of a future, nurtured by the anything-goes atmosphere of the day. When the FIM stepped in and outlawed the multi-gear, multi-

In the early 1970s Yamaha brought in a smoothly-styled, smooth-running ohc 750 twin to back up the XS1 650 which was anything but smooth but had plenty of gutsy performance in the manner of the ancient British twin on which it was fashioned. The TX750 was fitted with contra-rotating weights at the crankshaft, which tamed parallel-twin vibes at the price of subduing performance to a tediously low level. And it was not over-reliable, either . . .

left
Yamaha were disappointed with the 'Omni-phase Balancer' (their words) 750 twin but determined to have a 750 on the stocks. This is the XS750 of 1973, a value-for-money dohc triple with shaft drive which brought hope to BMW-fanciers short of buying power.

above
Every Japanese company believes that a 250cc four-stroke twin is a necessary part of its market line up. Yamaha's offering belongs in the straightforward 'get you to work on time' category. Worthy; but a little dull. But when the XS250 was introduced, the 100 mph RD was around to provide kicks.

right
A big company can cater for minority tastes, as Yamaha showed in adding a 500 single to their line. Sales have caused hardly a ripple on the corporate graph, but three years after its first showing the 500 continues to be offered — and bought mainly by middle-aged motorcyclists in the UK nursing ever-green memories of long-dead singles such as the BSA Gold Star.

cylinder specials, it was the YD-inspired racer that carried on, developed and doubled-up, but still basically the same, into the 1980s. And for road machines too the YDS was sole begetter — of YRs 2 and 3, the RD as 250, 350 and 400, even the watercooled LC twins of 1980. All bear the stamp of the 1959 YDS.

From the mid-1970s Yamaha, along with Suzuki, have climbed on to the four-stroke bandwagon with overhead-cam-shaft (usually dohc) twins, threes and fours from 250 to 1100cc, some with four-valve heads, some having shaft drive, and most (the exceptions probably limited to the 250/400 twins, no fire-brands by any reckoning) displaying plenty of style and competence. And when it came to a final surprise to round off the 1980s Yamaha chose to do it with a vee-twin, cylinders in line with the frame, that brought a tear (fond remembrance? indignation?) to the eyes of Vincent owners the world over.

The vee-twin Yamahas arrived following market research in the USA where, it appears, a schizophrenic situation exists among some influential motorcyclists who find indigenous Harley-Davidson at once attractive and repellent. They want a vee-twin but cannot stomach the H-D. Yamaha, ever-obliging and proving themselves currently the most quick-moving of the Big Four in R and D, came up with a 750cc vee-twin with shaft drive; and then, mindful of the European market, promptly enlarged it to 920cc, when the shaft was abandoned in favour of chain, presumably to endear the bike to spanner-happy riders in France and the UK.

Yamaha's vee is worth attention because it illustrates the flexibility of the Japanese mind which, while revelling in technical virtuosity, on occasion apparently for its own sake but more often as part of a policy contributing to ever bigger sales and profits, can swing to embrace the very designs elbowed aside in the early days of the new motorcycling. It is as if 25 years of motorcycle manufacture have shown Yamaha that some traits in the motorcyclist's personality endure through generations: a liking for easy, low-revving power, for instance, and do-it-yourself maintenance. Not that motorcyclists of this persuasion will ever again be a majority among the new 'bikers'. Changing social mores, incomes and expectations — not least the revolutionary ways of the Japanese constructors themselves — have seen to that; but still there exists, apparently, a number of people with a taste for something other than the eternal line of parallel twins, fours and sixes, water-cooling and banshee exhausts. Yamaha appear to be the one Japanese maker aware of these few people and, more important, willing to make motorcycles of a sort to keep them happy. It is not, of course, anything to do

with altruism. Undoubtedly sales will run into thousands. Yamaha first reacted to the 'vintage' sect when they introduced the XS 1100 dohc four in 1977. Alongside this 130 mph 550 lb projectile, overshadowed but a lot cheaper, was another new offering, the XT500, a 497cc single so uncomplicated, so ordinary, that it might have been a leftover from one of Norton's new-season menus for the 1950s.

The Japanese combine alert business reactions with forward planning — an unbeatable combination for any large-

above
When the RD250 finally petered out in the late 1970s Yamaha took breath, waited a few months to turn the screw for the bereft . . . and then announced a watercooled two-stroke twin, the LC, in 250 and 350cc sizes. With the LC the 100 mph barrier has been broken at last by a production 250. Technology and engineering by Yamaha are impeccable; protests and concern by interested 'anti' bodies — including anxious mothers — have been considerable.

As the 1980s began Yamaha went 'vee-twin'. This is the 750, with shaft drive, intended mainly for the USA.

41

Yamaha's XT500, which was labelled an Enduro but was in fact more suited to the milder forms of trail riding.

scale manufacturer. The new twin has what appear to be two XT500 engines mounted on one crankcase. If they are not precisely XTs (and they are not) it is nonetheless reasonable to assume that many components of single and twin are near-identical, for streamlined production. Where the Vincent had cylinders angled at 50°, and Ducati favour 90°, Yamaha have settled on 75°, which keeps the engine reasonably compact and the

wheelbase on the brief side of 60 inches while allowing room for carburettors between the heads. The single-throw crank, running in ball bearings, has the connecting rods side by side, and hence a slight stagger for the pots. Cam chains for the sohc on each head are driven from both ends of the crankshaft, the rear cylinder from the lefthand end, with tension being maintained by automatic adjusters. For the first time in many years

The engine/gearbox unit forms part of the chassis, depending from a large backbone in Vincent fashion. The rear suspension too is not entirely unlike a Vincent's, although repetition of the trendy word 'monoshock', for the Yamaha setup, may obscure the connection. (An American reporter would go no further than ' . . . single-spring designs certainly aren't new but have not been used recently on street bikes, so Yamaha get credit for reinventing the idea'.) On the Yamaha the suspension unit combines conventional coil spring and air, and has adjustable damping over a range of 20 positions, with use of magnesium in a critical area keeping the damping rate consistent whether the unit is cold, or heated by use. The damping adjuster and air valve (giving 7 to 57 lpsi variations) are on the right, below the seat.

The 750 has rear shaft drive, as noted earlier, which appears to be neat and conventional; the 920's chain drive is rather more interesting. Like an enlarged MZ design, it has twin hardened-rubber covers for the runs of the chain, fitting front and rear into cast-aluminium casings around the gearbox and rear-wheel sprockets. Lubrication is by grease and the chain is helped to keep on the sprockets by moulded-in rubber guides on the inside of the covers. The 920 has bore and stroke of 92 x 69.2mm, the 720 has the same stroke and 83mm bore.

Apart from some rather advanced features such as 40mm CV Hitachi carburettors and electronic ignition, and more or less old hat three-disc braking and cast-aluminium wheels, the new Yamaha appears likely to excite little comment in any pre-1950s *concours*.

A peculiarly 'British' kind of four-stroke is the big single; Yamaha brought in their sohc 500 at the same time as the XS1100 and it has sold reasonably well. A specialist dealer, Alf Hagon of East London, has used the short-stroke engine in a frame of his own to make an impressive cross-country machine.

a Japanese crankcase (road-going variety) is split vertically. It will be interesting to see how well oil-tightness is maintained. Other sources of seepage may be the external piping running from crankcase to heads, a feature reminiscent, like the crankcase split, of 40-year-old European practice. In Yamaha's case the outside lines are said to be necessary to *prevent* leakage where the metal head gasket would interfere with internal oil ways.

Kawasaki

Kawasaki is at once the biggest and the smallest motorcycle producer in Japan. Biggest because Kawasaki Heavy Industries, employing more than 35,000, with diversified interests in steel, shipbuilding, aircraft, trains and buses, is ranked among the leading Japanese — hence world — companies, dwarfing even Honda Motor; smallest because indisputably this mighty concern makes fewer motorcycles than any of the concerns in the business in Japan.

The original Kawasaki — for the purposes of this chronology — founded a dockyard in Tokyo in 1878. His ships were part of the story of Japan's empire building in the last years of the 19th century. By 1907 Kawasaki were making locomotives and coaches and freight cars, and their steelworks were committed to bridge construction and other large-scale civil engineering: it was all part of Japan's belated industrialization. Expansion continued through the 1920s until, in 1937, the aircraft division was formed, later to turn out thousands of aircraft for the war effort. A conglomerate of this size, a Krupps of the Far East, could not be damaged to any serious extent by Japan's World War Two defeat.

Within a year or two of the war's end, in 1945, business was going strongly ahead. In only one part of its manifold enterprises could Kawasaki be said to have some excess capacity. Nobody needed airplanes. So the aircraft division turned to the motorcycle market,

Long after the 500 Mach 3 was discontinued, but not forgotten, Kawasaki were selling a two-stroke triple. This is the 1976 400, which had little of the fiery personality of the 500 but at least retained the triple's appealing smoothness. There was, as well, a 250-3 in the range.

In 1966 Kawasaki introduced the two-fifty twin two-stroke Samurai and followed that within a few months by an enlarged version called the Avenger. The 62 x 56mm, 338cc engine had disc valves and gave 40 bhp to propel the 330lb Avenger at up to 110 mph. The picture was taken at an Earls Court, London, show.

left
Kawasaki, like the other makers, believe that medium-capacity four-stroke twins are important. Kawasaki's offerings — in both 400 and 250cc sizes — appear much like the equivalent Yamahas, and both makes trail the more sporting Suzukis. None, however, approaches the integrated good looks and sales record of the Super Dream Hondas.

below
Another view of the Gyoichi Inamura-designed Z1. Superb 'bulletproof' power plant and transmission, doubtful handling and near-130 mph top speed (on 2-star petrol) made up an irresistible package.

above
Read-out on exhaust emission for a Z1, the dohc 900-4 that Kawasaki introduced in 1972.

opposite
After the strong meat of their 500 and 750 triples, Kawasaki lapsed into circumspect (two-stroke) middle age with a 400-3 that was well-mannered, not very fast, not even very noisy.

producing transmission systems and gearboxes on its up-to-date assembly lines for the small two-wheeler makers. In 1950 a four-stroke ohv 148cc single-cylinder engine was designed and built — the first Kawasaki engine — and sold to Gusuden and Fiji, among the independents. Four years later a Kawasaki subsidiary, Shim Meihatsu, was selling machines fitted with a Kawasaki 50cc two-stroke. It should be understood that motorcycle interest on Kawasaki's part was at best sporadic — certainly compared with Kawasaki's attention to its other industrial efforts — and enjoyed only thin support in the boardroom. The 50, unimpressive to start with, was not developed and thus sold very slowly in competition with the Honda step-through and small Suzukis. Sales crept up, and then began to fall back in 1959 from the best annual total of 10,000 units, at a time when Honda were counting sales of the Super Cub in hundreds of thousands.

right
Derivatives of the Z1 four
included the 64 bhp Z650 of
1976, which had Honda 750-
level performance and better
handling (mainly because of
lower weight) than the big
Kawasaki. Unlike the 900 (but
like the Honda), it has plain
main bearings.

below
A 90 bhp 1000 (and a bit:
1015cc), the Z1-R of early 1978
had good looks but came over
as something of a 'cosmetic
job' to prolong the life of the
Z1-based four. Perhaps it
suffered by comparison with
the newly designed Honda
900F series and Yamaha's and
Suzuki's more modern
offerings.

What kept Kawasaki interested in making motorcyles, despite evidence of profound public indifference to their machines, was the notion that motorcycling was an effective way to put Kawasaki across to the world. Honda, of course, was the example that spurred them on. Decision time came at the end of the 1950s, when Meihatsu was slipping into ever deeper obscurity. In 1960 a plant was opened at Kobe to mass-produce motorcycles. The decision had been made. The first complete Kawasaki motorcycle was produced in 1961 — a 125cc two-stroke putting out 8 bhp at 6,500 rpm — and an organization was set up at Kobe to promote and sell it to an uneager world. Then within months Kawasaki, fired by a new all-or-nothing spirit for motorcycling, took over Meguro, the oldest motorcycle manufacturer still in business in Japan, at one time number two to Honda in the sales charts. Incomparably more experienced than Kawasaki, with an extensive range of motorcycles from 50cc to an unnervingly close copy of the BSA A7 twin, Meguro was happy about the alliance. Their motorcycles did not

resemble British offerings only in technical detail: though of excellent quality, they were beginning to suffer, in the same way as the British, from a dated air which could not stand up to the aggressive modernism of Honda and the others. But with Kawasaki's plentiful finances to tide the new partnership through a difficult year or two, there would be opportunity to design new models for more effective competition in the market.

In November 1962 the first motorcycle to bear the Kawasaki name was put on the road. It was an undistinguished 125, a development of the earlier model, and was named B8. Where other makers had followed the road-race path to instant publicity and increased sales, Kawasaki were handicapped in not having a high-revving ohc multi, or a 100 mph YDS, on the stocks; instead there was only the unpromising B8. (The 500 A1 obviously was no match for any racing 500.) Realizing that competition was not so fierce in moto-cross, Kawasaki did a Cinderella job with the B8, turning it into the B8M, a cross-country rocket. In 1963 it

Most of the latest Kawasakis are equipped with cast-aluminium wheels, like this ST1000, which was Kawasaki's almost obligatory move in turning a best-selling chain-drive sportster into a shaftie. Economics, for an owner, of chain v shaft over an extended mileage are unclear. Chains last longer than in the past, even with greater power on tap, because of improved technology; but cost a great deal more than before. Probably a shaft is cheaper.

49

took the first six places in the Japan Moto-cross Championships, and Kawasaki had their first competition success.

By 1965 Kawasaki were producing almost 30,000 units a year and decided it was time to go for exports — to the USA, naturally. Starting in Chicago, then shifting to Los Angeles, Kawasaki had a base from which to sell their two-stroke twin Samurai, a rotary-valve 250 that put out 30 bhp and was on a level footing with Honda four-stroke twins and Yamaha's two-strokes. The Samurai was later enlarged to 350cc, when it was known as the Avenger. The Americans liked both — to the extent that Kawasaki production more than doubled in 18 months from 1966 to '67. Satisfied with their showing in California, Kawasaki opened in New York, and for the occasion took an enlarged version of the 500 Meguro to America. The story of the 650 Commander's failure and Kawasaki's subsequent decision to build the export-only triple is told elsewhere. Suffice to say here that the 60 bhp Mach 3 was a major part of Kawasaki's final, successful operation in America, which by the early 1970s was disposing of 200,000 units a

year. In 1969 Kawasaki Heavy Industries was created by a merger among the many elements of the corporation, and there-after more direction and drive were diverted towards the motorcycle section. The 900cc Kawasaki dohc four, developed over an extra year or two when Honda's CB750 stole a march on Kawasaki by appearing 'early', came on to the market in 1972 and almost immediately upstaged the Honda in everything but sales volume. It won countless races, kept 'special' builders happy in several continents, was successively enlarged and detuned, then power-boosted, had a new frame, shaft drive — while at all times remaining very close to the original concept and maintaining Kawasaki's reputation as a manufacturer of fast motorcycles.

At the end of the 1970s Kawasaki added to that reputation by building another very fast, even larger motorcycle, the in-line six-cylinder water-cooled Z1300, which managed to dim the impact of Honda's CBX and, representing as it does up-to-date Japanese technology, deserves a more detailed analysis, which is to be found in the following chapter.

above
The first Z1 of 903cc put Kawasaki firmly at the top of the performance pile. One of the descendents of Z1 was the 1000cc Z1-R, which was a good looker but, somehow, deficient in the personality that had made the 903 so popular.

opposite
When Honda elected to withdraw the CB400 from the market (replacing it with a cheaper — to make: not to sell — twin), Kawasaki stepped in with their four. Nicely made, with dohc, having a good turn of speed . . . really, it was a rather better proposition than the Honda . . . the 400J has been singled out by one of the specialist papers as 'boring'. Which seems rather unfair.

51

Japanese Technology Observed:
Kawasaki Z1300 six

The Z1300 power unit on its own weighs only 50 lb less than Kawasaki's Z250 complete (286 lb against 336 lb), is 25¼ inches wide across the crankcase and is installed in a machine with a dry weight of 635 lb. Add some six gallons of fuel, a gallon of oil and three-quarters of a gallon of coolant, and kerb weight rises to well over 700 lb. With rider and passenger, weight approaches half a ton.

The first item of interest is that the machine is constructed to three different specifications, all powered by the same basic 1,286cc six-cylinder power unit but offering variations in riding position, fuel-tank capacity, emission control and power output. For all markets except Germany the engine produces 120 bhp at 8,000 rpm and a hefty 85.3 lb ft of torque at 6,500 rpm. However, owing to the theoretically voluntary, but in reality mandatory, German restriction on machines producing more than 100 brake horsepower, the version for Germany is equipped with a 99 bhp motor and accordingly generates a lower torque figure of 75.2 lb ft. The only measures taken to achieve this reduction in power are replacing the air-filter element with one having restricted air ingress through two holes, and having main jets two sizes smaller. All other aspects of the machine are unchanged, so it is understandable why this is a popular motorcycle in Germany, if only for ease in re-converting to the higher-powered version!

In fuel-tank size North Americans suffer with a mere 4.71 imperial gallons (21.4 litres) whereas the European versions are equipped with a 5.95 gallon (27 litre) tank. As most of the additional tank space is achieved by increased height, the two versions can be easily differentiated; the throttle cables sweep upwards, to prevent fouling, on the large-tank version and more conventionally

downwards on the United States and Canadian models.

The final major distinguishing feature is on the US model, which is equipped with a unique form of emission control — air injection into the exhaust ports, but without any form of independent external pump. The system is mainly located in the cam-box cover and air-filter box and is described later.

What does a customer obtain for his money? First — the engine, although exhibiting no radically new features, combines many unusual points in one refined power plant. Of 71mm stroke, 62mm bore, it is well under square, which helps to keep width to a minimum and assists in lowering emissions and fuel consumption by presenting a better surface area to volume ratio in the combustion chambers and more compact shape. Additionally water-cooling, with wet cylinder liners, assists in close

above
Water-cooled, six cylinders, double overhead camshafts, 120 bhp, 130 mph: the ultimate superbike?

opposite
From some angles the Z1300, while never lissom, manages to convey an impression of less bulk than hard facts insist. Here, for instance, it might well be a mere 500lb 1000 instead of the third-of-a-ton monster it is.

Superb power unit is remarkably compact, in view of its large capacity.

spacing of the cylinder bores and promotes more even running temperatures to give more consistent engine performance, particularly at the high engine speed and heavy loads which are the engine's *métier*. Piston clearances are a fairly tight 0.001 to 0.002 inches and the pistons are conventional three-ring items with two plain rings and one oil-control ring.

The radiator is mounted immediately in front of the aluminium block and connected in sealed manner to an overflow tank mounted on the rear lefthand side of the engine (although early design sketches were produced for the Z1300 with the radiator mounted in the nose of an integrally conceived fairing). Protection is afforded the radiator by a thick rubber shroud and metal-gauze screening, to give some defence against stone attack. Refinement is added to the system by a very compact electric fan mounted between radiator and block which is thermostatically switched in at a coolant temperature of 207°F and wired to operate whether the engine is running or not. Thus if engine heat is still being fed into the coolant with the machine stationary the fan will cut in automatically to reduce temperature. A conventional pressurized radiator cap is fitted, set at 11 to 15 lb per square inch, with water circulation controlled by a water pump and thermostat for rapid warm-up.

The cylinder head features arrangements typical of any large Kawasaki four-stroke motor — two valves per cylinder, each fitted with valve stem oil seals, with head diameter of 34.5mm for the inlet and 29.5mm for the exhaust, double overhead camshafts, dual valve springs and shims mounted above inverted bucket followers, similar to Z1 thousand style. Combustion chambers are virtually hemispherical; these, in conjunction with the well domed pistons, give 9.91:1 compression ratio but permit running on regular-grade petrol. The camshafts run direct in the cylinder head with four bearings per shaft and are hollow, for lightness. Drive to the camshafts is by narrow Morse Hy-Vo chain through integral cam sprockets. Valve timing is 20-70-70-30, in conventional nomenclature, giving a slightly later opening inlet than the earlier Z1 series, and valve lift is 8mm for the inlet side and 7.5mm for the exhaust.

Although the basic head castings are the same for all versions, the United States model has further machining operations performed which open up a normally blanked off passageway between each exhaust port and a cavity immediately underneath the cam-box cover. Each cylinder has a unique cavity which is covered by a reed valve mounted in the cam-box cover opening towards the exhaust port. The reed valves are in

Sectioned cylinder block shows fan, water pump, radiator and water passages.

Underside of the cylinder head.

turn connected to pipes feeding to a vacuum switch valve and from there to the air-cleaner box. Outwardly the system resembles the closed-circuit breathing system fitted to many cars but, although this latter system is fitted separately to the Z1300, the function of the various pipes and valves is simply to pass air direct from the air cleaner to the exhaust port on acceleration, because of differential pressures existing in these two areas. In the exhaust port the additional oxygen injected is sufficient to complete the combustion of unburnt mixture as well as oxidizing carbon monoxide to carbon dioxide, so reducing unburnt hydro-carbon and carbon monoxide emissions. On engine braking the vacuum switch closes and prevents passage of air and back-firing in the exhaust pipes; and the reed valves mounted in the cam-box cover prevent any blow-back from the exhaust pipe into the air cleaner.

The beauty of the system is two-fold in that it requires no additional motors or pumps, as is normal motor car emission-control practice; and secondly, the carburation can be jetted with regard to power output rather than emission control. Hence warm up, throttle response and general running are excellent with only a slight weight penalty over non-emission control models but no power loss.

Carburettors are three twin-choke 32mm constant-vacuum Mikunis of Solex diaphragm type. Although twin-choke carburettors are not new in motorcycle practice (both the Suzuki RE-5 rotary and the Friedl Munsch Mammoth have employed them), they form a particularly compact system giving a significant reduction in weight of fixed individual units. The carburettors feed into inlet tracts cast integrally with the head, the outers being of particularly unequal length — an arrangement which is bad in theory but probably functions perfectly well in practice. Fuel is fed via an electrically operated solenoid valve, as in Italian Morini and Moto Guzzi practice, from the tank, and air-filtration arrangements are catered for by a replaceable pleated paper element. The exhaust pipe of each cylinder is equipped with a blanking screw a short way down from the exhaust port to serve as an emission-control probe point, and the three pipes each side feed into a collector box and thence into a long silencer.

The most complicated and technically interesting area of the unit is undoubtedly the lower end, which incorporates a total of no fewer than three transmission shock absorbers, a torsional vibrational damper, four chains, an 18-plate clutch and a plethora of ball, roller, plain and needle-roller bearings and gears. The crankshaft performs a minimum of functions, with only an alternating-current generator mounted on its righthand end, and a conventional type of Japanese roller starter clutch driven by the starter motor through reduction gears on its lefthand end. It is a seven plain main-bearing item with plain big-ends and diagonally split, serrated connecting rods which can be removed upwards through the narrow bores. The drive is taken off the centre by a 34mm-wide Morse Hy-Vo chain passing round a 24-tooth sprocket integral with the shaft. Mounted on the extreme lefthand end of the crankshaft is a device rarely, if ever, seen on a motorcycle before the Z1300 — a torsional vibration damper. Why the need for a vibration damper? Because although a six-cylinder motor can be made exceedingly smooth in respect of out-of-balance forces, it is susceptible to torsional (twisting) vibrations. The torsional damper fitted comprises a small metal flywheel bolted to the end of the crankshaft, with its periphery rubber coated and an annular metal ring bonded to it. Hence vibrations are damped out due to the inner and outer members becoming displaced relative to each other when excited by vibration, and the exciting energy is dissipated as heat in the rubber coupling. Note that this device should not be confused with a Lanchester damper, which is a friction device. The Z1300 method of working can best be described as being very roughly equivalent to that of the friction clutch mounted in the drive gear of an old Lucas Magdyno.

From the crankshaft, drive passes to a jack shaft mounted immediately behind the crankshaft, which drives three further chains. The first two are of small size, one a conventional rollerless chain and the other a 9mm Hy-Vo cam chain. Also mounted at one end of the jack shaft is a nylon oil-pump drive gear and at the opposite end, driven through a face-cam type, single coil spring shock-absorber, is a further 40mm wide Hy-Vo chain transmitting drive to the clutch.

The small rollerless chain drives a further small jack shaft mounted above it in the rear of the cylinder block casting; this latter shaft is equipped with a bevel gear, which turns the drive through 90° and powers the water pump, and a small gear at one end to drive the ignition unit.

Carburettor intakes.

Both smaller chains are equipped with tensioners, the cam chain being automatically adjusted by a pressuring spring prevented from returning by a ball-race bearing on a gradual taper. Two guides are also provided for the cam chain and, as there is a reduction between the crankshaft and jack shaft of 24:32, the small Morse sprocket on the jack shaft and top camshaft sprockets restore the overall drive ratio to 1:2, giving the necessary half-engine-speed reduction for the camshafts. Careful selection of material and components ensures that these varied drive systems remain quiet in operation as one half of each pair of oil pump, water pump and ignition unit drive gears are nylon, with the matching half in steel; and Morse Hy-Vo chains are inherently quiet when running and long-lived, given adequate lubrication.

Resuming the devious power flowpath ... the clutch, rotated by the driven Hy-Vo chain, features a further shock absorber, this one of rubber-vane type. The clutch must be one of the biggest, if not the biggest, ever to appear on a motorcycle, having no fewer than 18 plates and an external diameter of 7½ inches; its torque capacity is immense. The five-speed gear cluster is mounted on two shafts with three selector forks, in conventional Japanese style, but is massive in the extreme. Output is then carried through a further face-cam shock absorber, this time controlled by four Belleville (dished) washers laid alternately (and functioning like a Norton Commando diaphragm clutch spring) and then through the hollow gear shaft to the first of the transmission bevel gears, through 90° and out via the mating gear. These bevel gears are supported on a needle roller and double ball bearing apiece and are lubricated by engine/gearbox oil.

The output shaft then feeds into the drive shaft via an internally splined, fixed four-bolt coupling and one universal joint to a splined coupling, to accommodate suspension movement, and then into the final-drive housing. The pinion gear here is mounted on two taper roller and one needle roller bearing, and the ring gear on one needle roller and a large-diameter ball bearing. Finally, output is threaded through a further rubber-vane type shock absorber to the rear wheel.

Shims appear to have been a design fetish. Apart from valve clearance, for which 25 sizes of adjustment shim are available, the gearbox output shaft shock absorber is shimmed in three separate places, the front bevel gear set in three

Sectioned US-specification power unit.

separate places also with seven sizes of each shim being available, and the rear bevel gear set has three separate types of shim, two of which are available in seven different sizes and the remaining one in 14 different sizes! All in all, a system of adjusting components which throws a rather heavy commitment on parts-stocking for correct fits.

Lubrication arrangements are very well contrived, with a trochoid type oil pump mounted in a sump containing 4.6 litres of oil. Oil is passed through a metal gauze to the pump and then via a replaceable oil-filter element, the housing of which is equipped with a by-pass valve, in case of clogging of the element, and pressure relief valve. The oil then follows one of five routes before returning to the sump under gravity. The main passageway is to main bearing journals numbers 1,3,5 and 7 and thence by cross drillings to big-end bearing journals numbers 1,2/3, 4/5 and 6 respectively and appropriate cylinder areas. Main journal number 1 also bleeds off to the starter clutch gear bearing, and numbers 2 and 6, apart from lubricating the mains, feed up to camshaft journals 1/2 and 3/4 respectively, on both inlet and exhaust sides, and from there to tappets, and cam lobes, returning via the camshaft tunnel, lubricating the timing chain en route. The middle main journal feed also lubricates the water pump drive-shaft bushes, ignition unit drive-shaft bushes, timing advancer shaft and, by directed oil jet, the primary Hy-Vo chain. The second passageway simply feeds the main jack shaft, the other jack shaft journal being fed by the third route via the oil pressure switch.

Numbers four and five passages feed each gear shaft, with cross drillings to supply bearings and gears as appropriate, including the bevel gear set contained within the crankcase. The rear bevel-drive unit contains hypoid gear oil, to either SAE 80 or SAE 90 viscosity, though the specification called for is to API-GL5, a higher lubricant additive specification than most conventional gear oils possess.

The electrical arrangements seem equally well thought out, with a 'pointless' ignition unit mounted behind the cylinder block driven by the previously mentioned jack shaft via a conventional mechanical/spring automatic timing device. The ignition unit comprises three pick-up coils and is simply a magnetic replacement for conventional mechanical contact-breaker points activated by timing rotor. The voltage pulses generated by the pick-up coils are then fed to what Kawasaki describe as an 'IC Igniter' which is, apparently, a Darlington power transistor mounted behind the righthand side cover and performing the switching operation on conventional double-ended ignition coils. The system is 'idle spark', i.e. it fires every revolution and employs resistive 14mm sparking plugs for suppression of radio interference. Adjustment is provided for in the system by rotation of the back plate on which the coils are mounted, although it is not possible to alter their relative positions, as the mechanical advance-retard unit can wear, for example, which stroboscopic checking of the ignition timing would reveal. Ignition advance range is from 10° BTDC at 1,050 rpm to 38° BTDC at 2,900 rpm.

The generator is wound in three-phase Y-fashion with a rated output of 224 watts at 5,000 rpm and a nominal voltage of 14 volts, the stator being of internal type and mounted in the outer cover with a rotor of dished construction to encompass the stator. Control is effected by a solid-state combined regulator/rectifier unit also mounted behind the righthand side cover and feeding to a 12 volt 20 ampere-hour battery. Three main fuses are fitted in the various circuits.

Instruments and other electrical accessories are very much in the Kawasaki idiom. All-new instruments and switch gear were introduced, with a large 60/55 watt quartz-halogen rectangular headlight and large two-bulb tail light. Switch gear features a rotating ring dip-switch, of similar style to that available in the 1930s, mounted on the lefthand switch console together with headlight flasher, indicator switch, hazard warning switch, horn and manual/non-manual cancelling indicator selector switches. This latter system functions in a manner similar to that devised for the Kawasaki Z1-R in that the flashing span is four seconds plus 50 metres, so the system adapts splendidly to both high-speed and stop-start conditions, allowing the most relevant item of time or distance to predominate. A clutch interlock, again a Kawasaki feature, prevents starting unless the clutch is pulled in first; no kick-starter is fitted. On the righthand switch console are fitted cut-out switch, light switch, and starter button plus front-brake stop-light switch.

The main instrument panel incorporates an 11,000 rpm tachometer, red-lined at 8,000 rpm, fuel level and water temperature gauges and 160 mph speedo. Warning lights are provided for neutral,

opposite, top
Bottom end: Hy-Vo drive from the crankshaft, jackshaft Hy-Vo cam chain, nylon oil-pump drive gear.

opposite, bottom
Crankshaft is fitted with alternator rotor, starter clutch gear and torsional damper.

59

Bevel gear set in the crankcase.

nate, with a top spine of approximately 1¾ inch diameter and lower rails of 1⅜ inch diameter. The steering head is heavily gusseted and incorporates taper roller bearings; the pivoted fork is massive and is mounted on taper rollers supported on a substantial forged member at the junction of three rear frame tubes. Front forks are of leading-axle type, to release space that would otherwise be taken up by the front-wheel spindle, to accommodate damper/spring internals and increase travel. The wheel spindle is supported by twin pinch bolts each side and the fork stanchions are almost 1⅝ inches in diameter, again double pinch bolt-mounted on the lower side. Internals comprise fork springs rated at 47.5 lb per in, plus spring caps offering the facility to 'fine tune' the forks between air pressures of 7.1 and 10 lb per sq in. Although this range does not seem a great deal, the beauty of the system is that the forks are converted from a linear spring rate to a variable and exponential one over the normal travel of the forks of 7.9 inches. The rear units are conventional spring/oil type and offer five pre-load settings for the dual-rate 72.6/100 lb per in springs.

Wheels, whose original styling was first seen on the Z1-R, are of standard Kawasaki seven-spoke cast-alloy pattern, of great strength, with 17 inch rear and 18 inch front diameters. Rim sections are of tubeless-tyre-suitable profile, of 3.00 and 2.15 inches width respectively, and equipped with 130/90 V17 (130mm width, 90 per cent aspect ratio) rear and 110/90 V18 front tubeless tyres of Japanese Dunlop manufacture.

Brakes feature Kawasaki's sintered-metal brake pads, similar in design to those by Dunlop in the United Kingdom, and operate on twin discs at the front of 11.9 inches effective diameter and a single rear disc of 10.8 inches effective diameter. The stainless iron discs are asymmetrically drilled, not for improved wet weather performance (a myth dispelled by the Road Research Laboratory), but to damp down any resonant vibration caused by the metal brake pads. Observations of Kawasakis fitted with these new metal pads shows what intimate contact they must make with the brake discs: overnight rust forms on the discs as a result of their very effective purging. Caliper piston diameters are 42.58mm front and rear, with master cylinders of 15.87mm, both standard Kawasaki figures, although the hydraulic pistons fitted incorporate miniature brake pads at the caliper end to prevent heat transference through the metal pad and piston into the brake fluid.

high beam, oil pressure, turn signal (for both sides) and hazard-warning lights. Incorporated in the ignition switch are steering-lock and parking-light positions, but in addition a further switch is provided on the lefthand side panel which is intended to isolate readily accessible power sources behind a metal plate but leave the ignition switch in the most convenient position. Its functions are a duplicate of the main ignition key positions except that the steering-lock function is replaced by that of seat lock.

Obviously the sheer mass of engine and equipment requires a fairly substantial frame, and this is most certainly provided. It is of double-cradle construction, and large-diameter tubes predomi-

General styling of the machine follows a recent Kawasaki theme in adopting fairly angular 'parallelogram' lines, but coupled with large rounded sections. A remarkable achievement of stylist Kurishma has been the integration of tank, power unit and radiator to disguise their extreme width. Aluminium-alloy master cylinders incorporate sight glasses to determine fluid level, and pillion footrests are mounted on polished-alloy castings, all of particularly neat design and finish. The dualseat has a pronounced step up to the passenger portion of the accommodation and lies at almost 32 inches from the ground, pivoting upwards for access to a tail compartment.

All in all, this is a highly impressive machine, though there may be doubts as to its ability to deliver consistent performance, in view of its obvious complexities. Yet, as previously mentioned, none of the technology employed is particularly innovative, when considered in relation to car technology as well as the

motorcycle world. In terms of specific power output the engine is very little different from the earliest of Japanese multis, despite the unit's far greater sophistication. For instance, the early Honda CB750 produced 67 bhp from 736cc (91 bhp per litre) and the original Z1 Kawasaki 82 bhp from 903cc (90.8 bhp per litre). Kawasaki's 1300, with its 120 bhp from 1,286cc, is very much on a par at 93.2 bhp per litre and may be expected to have the same enviable reliability record as these two former models.

A more valid query may arise from the opening remarks, in that a kerb weight of 700 lb plus can hardly be deemed an ergonomic figure. Who would have imagined at the time of introduction of the Honda CB750 that the Japanese would continue to develop ever larger machines, so that whereas in 1969 the CB750 was regarded as vast, by today's standards it ranks as a relatively compact, 'manageable' machine.

Advanced emission-control system.

Sport and Competition

Road Racing to 1968

Japanese involvement in international road-racing dates from Honda's entry for the 1959 Lightweight (125cc) TT in the Isle of Man. The RC142 twins assembled in Douglas on that historic occasion had four-valve heads, double overhead camshafts, a six-speed gearbox, and were said to turn out 18.5 bhp at 14,000 rpm. The power rating was unremarkable, the rev band something of an eyebrow raiser. Compared with the leading MVs, and the Ducatis, even the unhandy-looking East German MZs, the Hondas were gawky and technically old fashioned. The forks were of leading-link design with a forest of struts radiating from the lower end to support the mudguard and take brake reaction. At the top the stanchions met a feeble-looking steering-head fabrication forward of the spine main frame.

The engine was set high, mainly because lubricant was carried in a sump under the engine. The twin Keihin carburettors had flat throttle slides instead of the more usual barrel-type. Altogether it was a strange-looking device, the only concession to conventional good looks being the long, smoothly tapering megaphone exhausts. As one of the TT reporters for *Motor Cycling* was later to write, 'Another popular fallacy is that the Japanese are copyists . . . the Honda racer is, without any doubt at all, basically original.'

The riders were Japanese. They had plenty of courage but little experience. Europeans looked at the Hondas on the few occasions when the Japanese mechanics relaxed their vigilance and left the bikes unshrouded and found them

Jim Redman of Rhodesia was captain of the Honda race team in the early 1960s. He is on the 350-4.

amusing, archaic and largely irrelevant. In the race they were quite outclassed. However, they trundled around reliably enough to finish sixth, seventh, eighth and eleventh. This was disappointing to Mr Soichiro Honda, who with his boundless enthusiasm and optimism would not have been surprised by an outright win, but it was good enough to take the manufacturer's team prize. It was, in fact, a very creditable performance. Honda's race people went back to Japan to carry out a little development work. More than a little.

The first racing Hondas to arrive in Europe were these 125 four-valves-per-cylinder twins with dohc driven by lefthand shaft and bevel: they were ridden in the 1959 Isle of Man 125 TT. The engine depended from a spine frame and the leading-link forks appeared rather flimsy. Many changes were made for 1960.

Australian Tom Phillis, the Honda team captain, takes his four through Quarter Bridge during the 1960 Isle of Man 250 TT. A few weeks later Phillis crashed at the Dutch TT, breaking a collar bone, and Jim Redman joined the Honda team. Phillis took the 125cc world championship in 1961 and remained team captain until his death following a crash in the 1962 Junior TT in the Isle of Man.

They were happy with the four-valve arrangement, which had advantages in 'breathing', in lower reciprocating weight, and in the way it facilitated a central position for the sparking plug and, consequently, a higher compression ratio and detonation-free running. Their ideas tended to run counter to progressive European thinking of the time which, for the 125cc class, favoured a single, a two-valve head and rpm strictly in four figures.

Honda developed the 125, then turned to a 250: what could be easier than putting together two 125s to make a four cylinder 250? So the RC160 was born, a 16-valve, dohc 44 x 41mm racer with camshaft drive by bevel shaft, this feature distinguishing the racers from road-going ohc machines, where normal Honda practice has been always to use chain. Power was given as 35 bhp/14,000 rpm and top speed as more than 120 mph. Again the forks were leading-link, rather like those on a 10-year-old Greeves: had Mr Honda's snoopers found their way to darkest Thundersley, home of Greeves, in the early 1950s? The engine remained, by racing standards, unfashionably high because of the oil-carrying sump. At the Japanese Asama races, in August 1959, the 250 Hondas were harried by a Yamaha twin giving away 15 bhp but not much in speed; finally the Yamaha cracked up and Hondas took the first five places. There was plenty of time before the Isle of Man TT. Honda's R and D department got to work again.

The 'Greeves' forks were thrown away; in their place went conventional telescopic units with a hydraulic steering damper. The cylinders were inclined and a new gear drive was fashioned for the overhead camshafts. Ignition was by battery and coil. The frame was made of tubes in place of the pressed, and box-section, arrangement.

In the Isle of Man in 1960 the new 250 fours finished fourth, fifth and sixth; and the 125s were sixth, seventh, eighth, ninth, tenth and nineteenth. An Austrailian, Bob Brown, rode the fourth-place 250, just behind the vastly experienced Tarquinio Provini on his Italian Morini single. Perhaps the Hondas were not, after all, irrelevant. One or two commentators thought Hondas might win a TT some day. Honda decided that he would win *next year's* TT; at least that is what he has been reported as saying, years later, in one of his infrequent expositions to the press. Within a few weeks of the Isle of Man TT Brown was killed, at the German Grand Prix, and other Honda riders, mainly Japanese, were injured. By year's end the best showing had been second place in the Ulster GP.

In 1961 new Keihins replaced the flat throttle-slide carburettors and a remote oil tank took the place of the integral sump, allowing the engine/gearbox unit to be lowered. The frame was further modified on the lines of the already classic McCandless Featherbed design for Nortons. The 125 that year was rated at 23 bhp and the 250, known in factory circles as RC162, at 45 bhp, which was enough to propel this under-300 lb four at up to 145 mph.

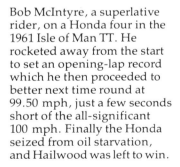

Bob McIntyre, a superlative rider, on a Honda four in the 1961 Isle of Man TT. He rocketed away from the start to set an opening-lap record which he then proceeded to better next time round at 99.50 mph, just a few seconds short of the all-significant 100 mph. Finally the Honda seized from oil starvation, and Hailwood was left to win.

A consistently successful but unspectacular rider, Jim Redman of Rhodesia tended to go no faster than was necessary in order to win — a policy which brought him innumerable first places, world championships and high earnings. More usually restricted, on Honda's behalf, to the leading GP circuits, he occasionally performed in the UK. Here he is at the Devil's Elbow at Mallory Park in October 1961.

Perhaps more important than technical improvements was Honda's decision to go into the market for riders from Europe and the British Commonwealth. The Japanese riders were charming, brave, and dedicated, when not racing, to acquiring a taste for European haute cusine — not least, in the Isle of Man, the celebrated Manx kippers. But they were out of their depth on the grand prix circuits, handicapped by too little of the 'right' sort of racing in their formative years. Honda signed up Tom Phillis and Jim Redman, Luigi Taveri, Tommy Robb, Bob McIntyre and, for occasional meetings, Mike Hailwood. At the 1961 TT Honda took first, second, third, fourth and fifth places in both 125 and 250cc races. Hailwood won both events, the 250 at 98.38 mph, faster than Phil Read's Junior (350cc)-winning speed and only 3 mph down on his own Senior-winning figure. (If it was a wonderful TT for the Honda Motor Company, then Hailwood had a delirious time.) The Scot Bob McIntyre, was a superb rider, utterly single-minded. He pushed his Honda to within .42 mph of the 100 mph lap and then the four, bedevilled by oil leaks, gave up and McIntyre had to retire.

At the end of the year Honda emerged as champions in two classes, Phillis taking the 125 and Hailwood the 250. The only flaw was that Phillis's title was in a sense his, and Honda's, by default following the abrupt withdrawal of MZ, the leaders, when their number one rider, Ernst Degner, chose to defect to the West at a critical stage in the series. But the record book was there to show Honda's

marvellous sweep of victories: ten for the 250 and eight for the 125.

At this point the beginning of a profound change in the pattern of European motorcycle racing was becoming apparent. Yamaha and Suzuki took a hand in the game. Suddenly it was no longer a case of the big men of racing, the Italians, getting their come-uppance from the clever Orientals; now the Orientals were squabbling among themselves and the Italians were standing sadly by, reduced to a supporting role. In a year or two they were to be — stretching the analogy — merely bit players. Accelerating the process, MV announced their withdrawal from lightweight-class racing.

Yamaha, of course, were committed to the two-stroke. In 1959 they had built a rotary-valve single which could reach 120 mph. Fumio Ito, the finest Japanese racing man to come to Europe in those early years, was equipped with a twin-

Surprise winner of the 1962 250 TT in the Isle of Man, Derek Minter rode a year-old four to beat the official works team of Tom Phillis and Jim Redman. He averaged 96.68 mph over six laps.

A 1961 Ulster GP shot, with three Honda men ahead of a lone MZ rider. Leading is Kunimitso Takahashi followed by Tom Phillis (19) and Jim Redman. It is the 125cc race.

two low-key seasons of European competition Suzuki had managed no better than a tenth place, in the 1961 250 TT, with New Zealander Hugh Anderson as rider. Usually the Suzukis, both 125 and 250, seized long before race-end. Degner concentrated on the 125 and a new 50cc racer. The 250 took something of a back seat. The best placing it achieved in 1962 was a fifth, in the Dutch TT, and then it was withdrawn for the remainder of the season. But the 50 did very well, with four wins and plenty of seconds and thirds, to take the newly introduced 50cc world championship — Suzuki's first. Degner was champion rider.

The 50 was an air-cooled single with disc inlet valve and expansion chamber under the engine. Power output was given as 9 bhp at 10,000 rpm and the gearbox was an eight-speeder in unit with the engine.

Over at Yamaha there was trouble with Ito's 250. Most of the following year was devoted to development work along Kaaden-Degner lines. In 1963 the 250 became the RD56 with anodized cylinders, hemispherical combustion chambers, enlarged exhaust ports and longer (50.7mm) stroke — all contributing to no less than 47 bhp at 13,000 rpm. Taking a page out of Honda's book, Yamaha redesigned the frame and suspension on the European pattern.

At the TT in June Ito finished second to Redman on the Honda; Tony Godfrey, riding for Yamaha, had opened with a lap at 95 mph, before crashing. In the speed trap Ito was timed at 145 mph. Yamaha's

cylinder development of the rotary-valve machine. The Yamaha arrangement had discs in the crankcase walls and a resonant exhaust system, following the precepts of the East German engineer Walter Kaaden, who had developed the MZ racing single to an astonishing 100 bhp/litre pitch in 1960/61. When Degner, with Suzuki connivance, defected to the West, he was soon on his way to Hamamatsu, where his engineering knowhow, hardly inferior to Kaaden's, was put at the disposal of Suzuki, and later of Yamaha.

Degner's job at Suzuki was two-fold: he was development engineer for the racers and number one rider. He set about rejuvenating the two-strokes. In

Ernst Degner in his first season for Suzuki, in 1961, took the world 50cc title. Here he sweeps the air-cooled disc-valve single through to victory in the Isle of Man. Power output was 10 bhp at 11,000 rpm which was enough, with an eight-speed gearbox, to push this 130lb midget along at close to 100 mph. This 1961 performance was the first TT win by a two-stroke since 1938 when the German, Ewald Kluge, took the 250 Lightweight on a DKW.

best showing of the year was a win at 120 mph on the very fast Belgian circuit at Spa, with Ito being backed up by another Japanese rider, Sumaka; the highest placed Honda was Tommy Robb's, at sixth. The season ended with another championship win for Redman — not that it was quite such a crushing win as in 1962; indeed the result was in doubt until the final round, at the Japanese GP at Suzuka. Redman also won the 350cc world championship in 1963, for the second time. The '350' Honda four was, in

a way, the brainchild of Bob McIntyre, who early in the 1962 season had pointed out to the race management that the 250 Hondas were a match for the European 350s. Boring out the 250cc RC163 by 3mm, to 47mm, was a simple matter for the factory. At 285cc the new Hondas were eligible to contest the long-established 350cc class. Known as the RC170, the 285 gave 49 bhp at 14,000 rpm. Successively enlarged to 49 x 48mm, 340cc, and turning out 50 bhp, and later to a full 349cc, the new four went through to the 1963 season

above, left
The first win by a Japanese in the Isle of Man was recorded by Mitsui Itoh in 1963. He rode a nine-speed Suzuki single to win the 50cc TT at 78.52 mph.

above
The Irish rider Tommy Robb was associated with Honda from as early as 1962, and rode for the company into the 1970s. This start-grid shot shows him on a 250 production twin in 1963.

The 1963 Honda 350-4, a classic design that netted Honda no fewer than four world championships in the 350cc class. Twin overhead camshafts and four valves per cylinder followed the layout that had been so successful on the smaller Hondas. Power was 52 bhp at 12,500 rpm. Only in 1967 was it pensioned off in favour of an enlarged version of the six that came out at 297cc and was too fast for Agostini's MV.

Ulsterman Ralph Bryans on the 1964 Honda 125-4. Power of the 35 x 32mm dohc unit was 25 bhp — enough to take Luigi Taveri, on a similar bike, to the world title in 1964.

for Redman's second class win. Indeed, the 350 was so fast and reliable that it won every 350 championship contested by Honda from 1962 to 1966.

But in 1962 Honda, and racing generally, lost two outstanding riders when Tom Phillis, the team leader, crashed and died in the Isle of Man TT and McIntyre was killed as a result of a race crash a few weeks later.

A new Honda of 50cc, introduced to the race scene in 1962, was a single with gear drive to overhead camshafts, four valves and an eight-speed gearbox built in unit with the engine. Power was put at 9.5 bhp at 14,000 rpm. It appeared to be no match for the Degner-inspired Suzuki and was the sole failure in the otherwise all-conquering Honda race effort of 1961-

1962-1963 (Anderson's 1963 125cc-category win for Suzuki excepted).

In 1963 it was Yamaha's turn to look for European riders. Ito was a skilled rider but business commitments in Tokyo and a crash in the Singapore GP took him out of racing. Yamaha invited Phil Read to take part in the Japanese GP at the end of the 1963 season. Read — brash, self-reliant, ex-Norton, ex-Scuderia Duke Gilera, a TT winner — flew to Japan and was beaten by Redman in the grand prix but showed, conclusively, that he was superior to any of the home-grown talent. He signed for Yamaha in 1964. Another team man was Canadian Mike Duff.

There was close racing, Honda v Yamaha, in 1964. At the French GP Read and Redman disputed the lead until the

Life in the paddock. A mid-1960s shot of Honda mechanics at the East German Grand Prix.

Honda dropped out with ignition failure, leaving Read and Yamaha to win. In the Belgian GP Read's engine locked and Redman, slipstreaming the Yamaha, had to take to the rough. Duff won.

At the Dutch TT . . . it may be enlightening to recall this meeting in some detail, to give the flavour of those 1960s battles between the Japanese camps. The Dutch race was the fifth round in the 1964 world championship series. It was the only major race meeting held in Holland, and the organizers devoted much attention to their job. A permanent staff was employed to run the event over a specially constructed 4.89-mile circuit. Highly popular with spectators and riders alike, the circuit embraced almost every conceivable type of bend, the only drawback being the lack of any perceptible gradients. Equal attention was given to spectator facilities, raised earthen banks with built-in seating covering the outside of virtually all the corners. The weather being so fair in Holland, covered grandstands were unknown — except for one, for the hard-working members of the press.

On race day in 1964 a vast throng — estimated by the organizers to be 120,000, possibly for tax purposes, in reality nearer 200,000 — crowded every vantage point. The weather was traditionally

'Dutch TT', almost opressively hot and with not a cloud to be seen — only droves of aeroplanes carrying advertising banners.

Nobody expected Redman on the Honda to have any difficulty dealing with Mike Hailwood (MV Agusta) in the first race, for 350s. In fact, he nearly did not have Hailwood to deal with because that gentleman arrived back at the circuit only an hour before racing began, having flown from practice for the French GP, for cars, at Rouen. From the start Redman rocketed away from the MV, and Hailwood never saw the Honda again. The previous year's Honda had been fast; but the 1964 version was rocket-like.

After his second lap, a record one, Redman eased off, but still put six to eight seconds between himself and Hailwood each lap. Hailwood held his lonely second place throughout, but a good fight for third had the crowd excited. From a poor start Remo Venturi forced his Italian Bianchi twin in front of Bruce Beale's 305cc Honda. But would Beale let the Italian go? He would not! He tucked in behind Venturi and tried all he knew to pass on cornering, but Venturi managed to hold him off. Both machines were timed at 133 mph — 6 mph slower than Redman's four but 4 mph faster than Hailwood's MV, which proved that

Well, it's running . . . Bertie Schneider, the Austrian rider, checks the exhaust of a Suzuki 250 in the Isle of Man in 1964. The square-four Suzuki, based on the 1963 championship-winning 125 twin (in effect it was two of the smaller units coupled, with water-cooling to prevent heating troubles with the rear cylinders) was fast but unreliable. It was heavy, handled erratically and, even when 'straightforward' carburation and ignition problems did not occur, was likely to seize. Frank Perris, who rode it throughout its short life, was glimpsed at a wet French GP sliding along the track, sitting up and, some distance from his capsized Suzuki, pushing his goggles on to his helmet with steely resignation. He had expected the square-four to seize, and of course, almost on cue, it had done so.

Hailwood was holding his second place on ability alone. Then, with half the race run, Beale's gearbox collapsed and Venturi was troubled no longer.

With only eight starters — four Suzukis, two Kreidlers, a lone Honda twin and a Derbi single — the 50cc race was rather farcical. That is, until the end of the first lap. Nose to tail, Hugh Anderson (Suzuki), Ralph Bryans (Honda) and Mitsuo Itoh (Suzuki) were having a tremendous fight. Bryans — a golden investment for Honda — was more than a match for champion Anderson and the two gradually drew away from Itoh. Then Anderson's ignition system broke down, and Bryans was free to win as he pleased. Itoh was joined by team mate Morishita, who had made a bad start, and they crossed the line together.

With Redman, for Honda, and Alan Shepherd (MZ) tying for the lead at the time in the 250cc championship, on 10 points each, and Phil Read (Yamaha) only two points behind, the 250 race was expected to be something of a match. An air of expectancy hung over the grid as the riders lined up. Read had been fastest in practice, but Redman was never in the habit of showing too much of his form before race day. In the mid-1960s he ranked with Luigi Taveri as the greatest tactician in the racing game. And then on the first lap, Read and Redman were tying for the lead. Shepherd had suffered a seizure on the MZ when lying second. And Provini? A bad start had left him near the tail-enders, and his Benelli four was down on power compared with the leaders' machines. Read's team mate, Duff, was holding a good third, with the Yamaha of Tommy Robb fourth; Derek Minter's Cotton was a creditable 10th, but

A 1964 250 Yamaha with 56 x 50.7mm engine developing approximately 48 bhp at 11,000 rpm, and seven-speed gearbox. Ridden by Phil Read, one of these RD56 models set the fastest lap in the Isle of Man 250 TT and went on to take the world championship, Yamaha's first. It weighed under 250lb and had a top speed of 145 mph.

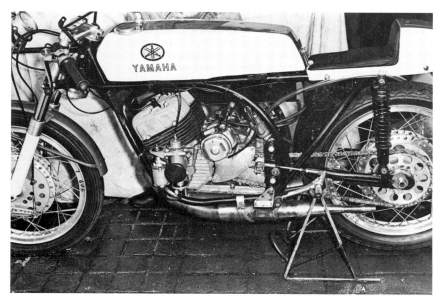

lasted only two laps before the big-end collapsed. The speed which Read's Yamaha had shown in the Isle of Man was proved to be no fluke. Out of some corners the Yamaha accelerated better than the Honda, but Redman was able to snatch back a few yards on other, faster, bends by dint of superior cornering technique. This was a distinguishing point about Redman: ordinarily a rider who went no faster than was absolutely necessary, he could, when needled, show himself as gifted as the best in the world. And Mr Read had the knack of needling his opponents . . .

Lap after relentless lap Read and Redman scrapped for the lead, roared on by the tremendously excited crowd. Read twice eased past Redman on the home straight, once dived inside him into a

series of S bends, but each time Redman snatched back the deficit on the twisty back leg of the course. Lost in their private battle, both were so far ahead of the field that soon they were lapping tailenders. By the finish only the third man was to remain on the same lap. Within four laps Duff, no mean performer (years later he was to say, in a burst of uncharacteristic spleen, 'I could have beaten Read, but of course I had to ride to team orders'), was 20 seconds astern and Bert Schneider on the Suzuki four had overtaken Robb, although the little Irishman was not inclined to give in so easily so early, and fought back. He was obviously regaining his old confidence and form, sadly depleted by his summary dismissal earlier in the season by the Honda management.

After eight laps, with Read and Redman still only inches — literally — apart, Duff stopped for a plug change, letting Robb into third place; Schneider had pulled in with some not entirely unexpected trouble with his Suzuki (of which more later). Duff restarted eighth, began the climb upward, eventually to claim a worthy fifth. Provini (to put the scene in perspective it is necessary to mention the activities of riders of non-Japanese machinery) had been methodically ploughing through the field until he reached fifth on Lap 6, then fourth, when Duff dropped back. But towards the end the Benelli partially tightened and Provini eased his pace, though without relinquishing fourth position.

Into the last lap Read had a slight lead, which he still held halfway round. Then

Hugh Anderson, four times a world champion, on his 125 Suzuki in 1968.

The best racing of a decade was seen in the mid-1960s when Jim Redman battled to keep Honda ahead of the improved Yamaha ridden by Phil Read. Early in 1964, Read's first year with Yamaha, it was plain that the Honda four had met its match. This shot at the West German GP shows the close racing that was typical of that year's meetings. Redman was ahead at this stage but had to give way to Read, in both the grand prix and the world championship.

. . . and a year or two later Hailwood was Honda's man and Read had Ivy to back him up. This is the French GP, and Hailwood was ahead of the two-stroke riders — until gearbox trouble slowed him and Ivy went on to win.

Redman made his bid and slipped past, only to make a nonsense of the next corner and let Read through. Into the penultimate left-hander; Read drifted wide and Redman went inside; then Read tried again on sheer acceleration out of the last corner. He drew level but failed by inches to push in front. A fitting tribute was the victor's laurel shared between the two on the rostrum. (Another tribute, years later, from the eloquent Read: 'The bastard! But you had to hand it to him. He certainly knew how to win races'.)

And then, astonishingly, the whole story was re-enacted in the 125cc race. That gives a measure of the race tempo of those days. The new 125 Yamaha was slower than Redman's four in practice, but Read must have been foxing. From the start he shot away to gain a clear 20-yard lead from Redman and the Suzukis

Finally, when Honda pulled out of the hunt, there was nothing left for Read and Ivy to do but fight among themselves.

of Anderson and Schneider; these three were nose to tail. But within a lap Redman had outspeeded the Suzukis and caught Read. Schneider found more power and managed to keep the two leaders in sight for half a dozen laps. Anderson dropped back to have a private war with Frank Perris (Suzuki). Already Shepherd on the MZ was out: another seizure.

As Read and Redman swapped places every lap, the second Yamaha twin, ridden by Robb, retired with plug trouble. And championship leader Taveri . . . where was he? Taveri had fallen in practice and hurt his head and shoulders: it was not serious but enough to keep him in bed for a week and out of racing until the West German GP, three weeks off. Another Swiss non-starter was the popular Roland Foll: he too had crashed a Honda in practice, but his injuries had proved fatal.

The Redman-Read duel distracted attention from Ralph Bryans on his Honda. Gradually he climbed until he had third man Schneider in sight. On the ninth lap he caught Schneider, and then began a fight for third which lasted to the finish, the pair changing places on almost every corner; both, though, were at least 18 seconds behind the two leaders. Then two laps from the end Redman managed to put 50 yards between himself and his shadow, and Read was unable to recover before the finish. The main excitement then was the Bryans-Schneider battle. On the last corner the young Belfastman managed to slip inside Schneider and take a yard lead to the end.

After the Dutch TT came the West German round, in which Read tucked in close behind Redman through most of the 250cc race, then out-accelerated the Honda to the finish line. In East Germany, at the Sachsenring, Mike Hailwood was performing on a 250 MZ to the great delight of 300,000 patriots and the deep concern of Read and Redman, who were split, Redman in the lead, by the dashing Hailwood. But not for long. Hailwood crashed on a bend and Read, slipstreaming, was lucky to avoid him; when he collected his wits, Read set off after Redman. In Read's own words: 'I sat behind (Redman) for a while, then it was all over as I asked the bike to go. We knew we had it over Honda'.

Honda knew it too: and even more forcibly when their man was beaten by Read in the Ulster Grand Prix. They had planned to introduce a six-cylinder 250 the following year, 1965, but Read's obvious superiority was putting the 250cc

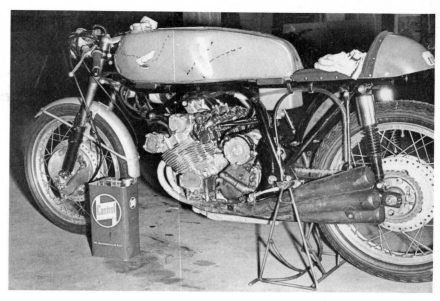

championship, for three years a Honda monopoly, in danger. The six would have to be brought into action earlier than intended: by not later than the penultimate round of the series, the Italian GP at Monza. To keep the crown Honda needed to win at Monza and at the final meeting of the year, the Japanese GP. The six was completed in great haste in Japan, with testing inevitably curtailed in the short time available. Then it was flown to Italy.

Read has recalled the scene at Monza, where the Honda was kept under wraps during most of the practice period. 'The first indication that there was something funny afoot came not long after arriving at Monza. Word was out that there was a six-cylinder Honda being kept under cover in the Redman camp. Stories like that circulated around GP paddocks all the time . . . But I must admit I felt a little worried. What was I to do? If I walked over to Redman's transporter, trying to look innocent, they'd have known straightaway they had me worried. Someone would have warned, "Here comes Read", and they would have sat around, drinking Coke and playing cards as if there was nothing better to do. And somewhere around would be this "thing" covered with a tarpaulin. I'd go away no better off and Redman would know he had me worried. So in the end I stayed away. He still had me worried, but he might not have known it. It was a war of nerves'.

In the race the six sped away from the Yamahas, springing a lead of 100 yards within the first minute. But by the seventh lap Read was closing on Redman though the Yamaha was going no faster; the Honda was in trouble and slowing,

Introduced at the end of the 1964 season, at the Italian GP, Honda's 250-6 superseded the four in an attempt to match the speed of the Yamaha two-strokes. Overheating slowed it at Monza, however, and it was only in 1965 that the 54 bhp eight-speed Six was able to demonstrate a convincing superiority over the Yamahas, though not frequently enough to prevent Read taking his second world title.

Mike Hailwood, arguably the finest motorcycle road-racer of the postwar years, on a 250-6 Honda. Hailwood, who retired as a world champion in 1967 and made a brief, storybook comeback 10 years on, when not far short of 40, died in a car crash in March 1981.

with the engine overheating and misfiring when Redman attempted full throttle. Read passed and was on his way to winning his, and Yamaha's, first world championship.

Battle lines in the 250 class in 1965 were blurred — to start with, at least. Redman (or Honda) elected to stay away from the opening grand prix, at Daytona; then he crashed at the second, at Nurburgring, which sidelined him there and at grand prix number three, at Barcelona. When he recovered to ride the six at the next GP, at Rouen, the Honda's gearbox locked up while he was establishing a 13-second lead over Read. All this meant that, as the teams went to the Isle of Man for the TT in June, Read and

Yamaha had won the first four grands prix. In Manxland Read struck bad luck in the 250 race, when he had to retire while holding the lead. But he won the 125cc race with a new water-cooled Yamaha twin, Yamaha's first Isle of Man victory.

Redman and the extraordinarily fast six were unbeatable on the wide open stretches of the Belgian and East German grand prix circuits, where the Honda clocked a higher race average than the winning 500 MV Agusta ridden by Mike Hailwood. A win for Read at the 10th GP, the Ulster, gave him his second 250cc world title. At the end of the season Honda was ahead in the 50cc class, where the 20,000 rpm twin had stolen a march on the Degner Suzuki; it was Honda

below
A 1964 125cc Isle of Man TT start scene. Number 12 is Hugh Anderson (Suzuki), number 11 is Luigi Taveri (Honda). Taveri took his four to victory not merely in the Island but in world rankings throughout the year, to wrest the world championship from Suzuki.

again in the 350cc class, with the four performing as reliably as ever (though, ominously, it had been harried on one occasion by a bored-out 250 Yamaha ridden by Read); Yamaha had taken the 250 title; and Suzuki was once more ahead in the 125 class in which Honda, in sharp contrast to their all-conquering display of the previous year, did not manage to take a single race.

That the six did not win the championship must have been a big blow for Honda's pride, but they had only themselves to blame. They were unfortunate in losing the services of a talented rider, Alan Shepherd, before he had even raced a six in support of Redman. But there were plenty of other

good riders ready and willing to be signed up. Honda were unwise, and unfair, in making Redman take all the responsibility, all the way. Much time and effort were being devoted to getting the Honda GP car into shape. For the six to succeed against the Yamahas it would have been necessary for Honda to have (a) another top-class rider to support Redman, (b) a more serious assault on all the classic races, with full factory back-up.

And Suzuki? Something has been said of their successful efforts in the 50 and 125cc classes, with Degner and Anderson as the leading riders. But Suzuki had ideas about 250s too, which took the form of a 43 x 42mm four arranged in square formation. The cylinders were water-cooled, with four carburettors feeding via four disc valves and three transfer ports into squishband heads. The firing arrangement had diagonally opposed cylinders sparked in unison, to give the effect of 180° twins mounted in tandem. Cooling water was circulated by thermo-syphon from a radiator mounted beneath the steering head and lubrication was by 20:1 petroil. This 50 bhp unit with six-speed gearbox was mounted in a conventional duplex-tube frame with telescopic forks and pivoted-fork rear suspension. Wheels were 18in, brakes of single-, then two-leading-shoe design, and the whole thing added up to 300 lb of

not very manageable or reliable motorcycle: as Schneider in the Dutch GP, and other riders, had cause to lament. In two seasons, ridden by people as various and talented as Frank Perris, Jack Ahearn, Schneider and Anderson, the square four's best placing was a third, in the 1965 Dutch.

One of the chief problems, apart from apparently inherent unreliability, was excessive weight, which led to a reduction in acceleration out of corners and too much speed into them. This lack of braking effect with the two-stroke fours was a major problem, and was to be satisfactorily resolved only with the introduction of disc braking, on which Suzuki were working. Undoubtedly the man of the season in the 125cc class was Hugh Anderson, the world champion. In spite of forays into moto-cross where he performed on Husqvarna and Metisse with great élan but only modest success (and eventually crashed, which caused him to miss the East German Grand Prix) Anderson secured the class for Suzuki with a brilliant performance in atrocious weather at the Italian GP in September.

At the start of the event only a couple of points separated him from the other Suzuki stalwart, Perris. Although it was known that the factory wanted Anderson to win the world title, it was known too that Anderson had not endeared himself to his Japanese masters by his extra-

Luigi Taveri on his 1965 16 bhp Honda 50 with 10-speed gearbox. With a top speed comfortably over 100 mph, Taveri won the Isle of Man 50cc TT, and Ralph Bryans, similarly mounted, took the world championship, despite the best efforts of Suzuki with their new water-cooled twins. The front brake is of bicycle caliper type, acting on the rim of the 18in wheel.

mural activities in moto-cross. Anderson had put himself a little more firmly in the dog-house at Brno during the Czech GP, when he relaxed concentration momentarily and allowed the Suzuki to clout a kerb, destroying several thousand pounds' worth of valuable machinery and letting Perris through to win. The factory were not particularly concerned that *Anderson* should win the title for them: Perris would do just as nicely. Then, at the Ulster, Perris retired with a broken crankshaft, leaving the third Suzuki, ridden by Degner, to win handsomely.

It was an occasion when Degner's extra two-stroke knowledge had been put to work. Before the race he realized that a strong tail wind on the main straight would cause over-revving in top gear, so he raised his ratios, thereby saving the engine. Both the other riders left their practice ratios unaltered and reaped the seemingly inevitable result of wrecked engines. So the tight little race for the world championship, albeit a one-make effort, moved to Finland where, because the country was behind the Iron Curtain, the defector, Degner, was unable to race. Degner was genuinely concerned that if he were to reappear within snatching distance of a Communist regime, a long arm might stretch out and put him back where he would, at that time, have least liked to be — the German Democratic Republic. Poor Degner! His hopes of

ultimate success in any but the 50cc class seemed doomed, for at least three race meetings in the championship series were held in Communist countries, with the 50s being summarily ignored; and then even the West began to lose interest in the 'babies', cutting them from the major grands prix, so Degner finally had little hope of succeeding even among the 50s.

In spite of all this irresolution in the 50cc class, with the European constructors falling out and only Honda and Suzuki left to dispute the title, some of the closest fighting in 1965 racing was seen among the tiddlers notably in Japan, at the final round. Here Bryans and Anderson, Honda v Suzuki, started level on points. Bryans finished second, to Taveri (Honda), and Anderson crashed trying to outpace them. So the title, for the first time, was Honda's.

In Finland, Anderson and Perris on 125s shook off the MZ challenge of the English rider, Derek Woodman, and settled to a personal fight until, gallantly, Perris moved over a couple of laps before the finish to let Anderson through. In Italy, as mentioned earlier, Anderson gave one of his finest performances when, despite falling off on the sodden track, he fought back to win by more than a lap from Perris. This splendid ride won him the title for the second successive year with, of course, Perris as runner up.

Honda extended their multi-cylinder doctrine for racers even to the 50cc class, with this 33 x 29.2mm twin of 1964 that revved to 19,000 rpm. A nine-speed gearbox helped to translate 15 bhp into 100 mph. Wheels were hardly bigger than those for a sturdy pedal cycle, at 2.00/2.25 x 18in. When its early troubles were sorted out, the twin went on to give Ralph Bryans a 1965 championship victory.

overleaf
Giacomo Agostini went from MV to ride for Yamaha. He is seen here in 1974 on a 750.

In the 125cc class, in 1965, it was Honda's turn to suffer more than a few slings and arrows, although none of their troubles could in fairness be attributed to 'bad luck'. Poor preparation, insufficient spares, lack of interest in Tokyo — these were some of the reasons why the four cylinder, unbeatable in 1964, never got into its stride. Lack of knowledge by the mechanics concerning the transistorized ignition system caused early setbacks, but even when the engines were seemingly going well they were outclassed by the Suzuki two-stroke twins. When something broke, or was bent, there were never enough spares on hand for repairs.

Honda's 125 riders, Bryans and Taveri, performed very well though they were frequently dispirited by the apathy of the factory. In truth the apathy was probably more imagined than real for, as mentioned earlier, the Honda experimental department was almost wholly engaged at the time on the racing car project; motorcycles were being put aside. Eventually, after the Dutch GP, the sole remaining four was declared no longer a fit runner and Honda abandoned the class for the remainder of the season until, in October, at the Japanese GP at Suzuka, they wheeled out a beautiful device with five cylinders. It was derived from the championship-winning 50: a simple matter — well, it was, apparently, simple for Honda — of multiplying ($\times 2\frac{1}{2}$) the 50's twin layout and ending up with a 33 x 29.2mm five with the exhaust pipe from the middle cylinder hoisted over the cam boxes and ending up under the seat. (Any recalcitrant team man could then be dismissed to take the hot seat of 125 racing.) The five-cylinder produced 30 bhp at 18,000 rpm, in later versions

going to 35 at 20,000, and at Suzuka in the hands of Taveri it looked as if the Honda would be too much for Anderson's Suzuki, until a broken cylinder-head bolt and subsequent oil leak slowed it to second, just ahead of Bryans' similar model.

It was not Suzuki alone who had worried mighty Honda through the season: Yamaha too had become a threat. Yamaha had brought out an air-cooled twin, the RA97, which was extremely fast but prone to seize up at inconvenient moments (when wasn't it inconvenient to have a seize-up in the race rider's busy round, you may ask?). A development of the RA97 substituted water for air as the cooling medium, when around 28 bhp was reliably available, as was demonstrated in the 125 TT in the Isle of Man. Read's lead — over Taveri on the Honda — in that race was eight seconds, and would have been considerably more had not a piston ring snapped during his final lap and slowed him a little. Yamaha's achievement was even more notable because reliable Duff brought the second water-cooled model into third place, only 15 seconds behind Taveri; the third Yamaha, an air-cooled RA97, came home seventh, which contributed to Yamaha's manufacturer's team award — their first in the Isle of Man.

They went on to a repeat win in the Dutch GP a week later. This time it was Duff's turn to win when Read dropped out with a repetition of the TT piston-ring failure. At which point, incredibly, instead of going all out to capture the world title in the class, Yamaha shipped the 125s home for 'further development', a move which was gratefully received by Suzuki. It was in the 125 Isle of Man TT

The 125 five-cylinder Honda in 1966. It had a top speed of 125 mph, achieved on about 30 bhp at 18,500 rpm. Taveri rode a five to his third 125cc world championship, in 1966.

that another bombshell startled dedicated race followers when the Suzuki reserve, Yoshimi Katayama, lay third on the opening lap behind the two leading Yamahas and in front of his renowned team mate, Perris. Not only that but he, in company with Read and Duff, had broken Taveri's 1964 Honda lap record from a standing start. It was a remarkable performance for someone who had never set foot on the Island until a couple of weeks before the race. Later, in Holland, he pushed Duff very hard in the 125 race and followed that by the best-ever Suzuki square four placing, at fourth, in the 250 race. A week later he again finished fourth, in the Belgian GP, well ahead of the experienced Perris.

By the end of the 1965 season Yamaha knew, despite Read's second win for them in the 250cc championship, that the ageing RD56, coaxed to give more than 55 bhp, was at the end of its run: the superb Honda six would be too much for

it in 1966. So at the Italian Grand Prix in September 1965, just a year after Redman had brought the six on the scene, Yamaha produced the RD05, a 58 bhp water-cooled 80° V-4 with four disc valves, eight-speed gearbox and a reputed top speed of 150 mph. It was in essence a pair of the very successful 125 engines bolted together, with upper and lower crank-shafts driving a common transmission gear. Two cylinders were vertical, two horizontal, and one crankcase was mounted above the other, all of which made for a high centre of gravity and consequent erratic handling. Read later rode the RD05, modified by way of rotary-valve changes and the installation of capacitor discharge ignition to produce more than 60 bhp, when the code name became RD05A, at the last meeting of the year, at Suzuka in Japan. Here he found it as fast as any 350 he had ridden but deficient in handling. In fact he fell off at 140 mph in practice, which resulted in

Phil Read came to Yamaha with a TT win qualification and proceeded to give the Japanese firm world titles in 1964 and '65. His straight-forward air-cooled twin had astonishing speed, though stamina was suspect; it took Hailwood on a Honda six to defeat him.

Bill Ivy being recalled at short notice from London to rejoin the team. Ivy finished third behind the two factory Hondas, one ridden by Mike Hailwood, the reigning 500cc champion, on an MV Agusta, who was in his first days as a contract rider for Honda.

Honda's preparations for 1966 were thorough. The six was going well, giving 60 bhp and having a wider, more useable power band than in its earlier form. The 125 was to be the five. The 350 four was developed to produce more than 70 bhp at 14,000 rpm. R and D engineers in Japan thought this would push the Honda ahead of the MV Agusta which had, in Giacomo Agostini's hands, run Redman's Honda very close through the 1965 season before the Rhodesian edged ahead to give Honda their fourth world championship. Finally, to keep Hailwood fully occupied during 1966, Honda wheeled out their first 500cc racer, a transverse 16-valve four driving through a six-speed gearbox and producing more than 80 bhp at 12,000 rpm. This remarkable motorcycle, throughout its life a clear leader in the class in terms of sheer horsepower, was developed within a year to give more than 110 bhp. At no time, however, was the engine mounted in a frame which enabled the power to reach the road without calling on every particle of Hailwood's skill and courage.

above
Read on the 250 V-4 Yamaha. Top speed of this complex water-cooled racer was around 150 mph.

left
Mike Hailwood's factory 500-4 racer restored to former glory 10 years after its heyday.

Honda, and Hailwood, had a tremendous time in 1966 despite the 500's frightening lack of stability, which prompted the R and D department to consider approaching Italian frame-building specialists. This was an unprecedented move for the Japanese, always sensitive to loss of face in international circles. Finally Hailwood persuaded Japan to let him commission a frame to be designed and built by the Reynolds Tube Company in England, but for most of the season he was to be seen wrestling with the standard factory racer. Very soon those riders, not a few of them, who had gasped on hearing of his £40,000 contract with Honda, were wishing him well and saying that possibly he was underpaid when it came to keeping the 'monster' pointing in the right direction. His antics were in marked contrast to the smooth, unruffled style of Agostini, his rival in the class, on the well-mannered MV-4.

Hailwood's team mate Redman, by seniority team captain for the Honda race effort, was experiencing something of the other side of the smiling face of the Japanese as employer. Honda had been enormously impressed by Hailwood's winning displays on his 350 MV and on an unfamiliar Honda 250 six at the end-of-season Suzuka. Redman, it was thought, had lost his fire and should take on a non-riding role with the team. Or at best step down to being number two to the new boy. But Redman did not consider that he was finished. In a personal confrontation with his masters in Japan — always a good tactic in a fraught situation for a non-Oriental, especially an obdurate man like Redman,

when the deviousness of the Japanese might be overridden or simply ignored — Redman secured factory support for his 500cc rides. As it turned out, he was the first to ride one of the 500s, in the opening classic of 1966, at the West German Grand Prix at Hockenheim. Hailwood, it was said, had been outwitted by the Rhodesian in behind-the-scenes race politics. He was down to ride in the 350 race against Agostini, which he did, duly beating the MV man. In the 500, Redman showed that he remained eminently employable by running away from the Italian. Quite clearly, the new Honda was too fast for the Italian bike, which had taken eight successive world championships.

Honda kept firmly ahead in all other classes in the early meetings of 1966, mainly through Hailwood's efforts. He was well clear of Redman in the 250 and 350 categories, though the latter, building on that first success in Germany, maintained his lead in the 500 class over Agostini up to and including the Dutch GP, despite the Italian's switch to a newly developed three-cylinder which was demonstrably a much better handler than the Honda and only marginally slower. At the next meeting, however, in Belgium, Redman crashed in the rain, breaking a collar bone. This meeting was effectively his last grand prix as a factory rider; within weeks, though he turned up for practice at the Ulster GP, he announced his retirement from racing.

Hailwood on another 500 had to drop out too, because of gearbox failure. From this point he took sole charge of Honda's race effort in 250, 350 and 500 classes. By

Honda, Hailwood and mechanics.

the end of the year he had won three 500 races, including the Isle of Man Senior TT, Honda's first victory in this event, with Agostini coming second on each occasion. But Agostini too ended the season with three wins, and with more points from early races he won the championship. In the manufacturer's category, however, it was Honda all the way, with wins in every solo category.

Read had not been able to deal with Hailwood in the 250cc class. Perhaps it would be fairer to say that the Yamaha, tall and heavy but possibly a match on power for the Honda, handled even worse than the six. Read was second to Hailwood in the Dutch, Belgian, East German and Czech GPs. So Yamaha, with Read, ended as runner-up among the 250s; and in the same position, by courtesy of Bill Ivy, in the 125cc class.

Though Honda, as manufacturer, took the 50cc title, the individual champion was Hans Georg Anscheidt, on a Suzuki, who had earlier decided that he had little hope of winning a world title should he remain with the German firm of Kreidler. Development of the Suzuki had taken power up to 16.5 bhp. Carburettors on the 130 lb water-cooled twin were updraught, with rear-facing exhaust ports. A single gear on a slave shaft behind the crankcase took power from the inboard end of each crankshaft; additionally, the shaft was connected to the magneto drive, and to oil pump, rev-counter, water pump and clutch. The all-indirect gearbox had 12 speeds — for some circuits, 14. This machine, ridden by Anscheidt, managed two further titles, in 1967 and 1968, though some kudos for

Suzuki was lost through Honda withdrawing from the class partway through 1967.

So 1966 was a Honda year. Because of the company's move into 500cc racing this class, which for some years had suffered from MV's near monopoly, with a consequent stagnation in technical development and fall-off in general interest, had been resurrected to its once unchallenged position at the head of grand prix affairs. As far as 125s and 250s were concerned, Yamaha had clearly taken a beating; equally evident was their determination to fight back in 1967. For this effort the 250 V-4 was further improved, with extra stiffening at steering head and rear-fork pivot to improve handling. In addition, two hydraulic steering dampers were fitted and provision was made for adjusting the steering-head angles: anything to make this 75 bhp nightmare more controllable. The water-cooled 125 twin was dropped

A Yamaha production racing 250, the TD1B of 1966, was based on a roadster design. The 246cc engine had 56 x 50mm dimensions and power output was 35 bhp at 10,000 rpm.

Yamaha's 125cc four appeared partway through the 1966 season and went on to a very successful 1967, giving Bill Ivy the world title. Cooling water was circulated by a pump on top of the gearbox, alongside the oil pump, with both pumps driven off the gearbox by spur gears. The gearbox had 10 ratios, the engine produced 41 bhp/ 17,000 rpm and maximum speed was at least 130 mph.

Chris Vincent, now fighting a claim for £100,000 awarded against him as a result of a sidecar crash, pictured in happier times when he was showing the way to many 500s-and-above with his T20 Suzuki in 1966 long-distance events.

in favour of a water-cooled V-4 that was a scaled-down version of the 250 RD05A, with the two upper cylinders canted forward at about 30°. Built in unit with the engine, a 10-speed gearbox transmitted 40 bhp at 18,000 rpm.

In 1967, in the 250 class, Read and Hailwood on the 62 bhp six battled for the title. The lead changed from Yamaha to Honda to Yamaha as the season progressed. Finally, at the Japanese GP in October, Honda and Hailwood beat the Yamaha-Read combination by adjudication on the basis of the points tally through the season. So Yamaha failed, just, to topple Honda in the 250. They were more successful with their 125s, winning the Isle of Man TT (Read), French GP (Ivy), Ulster GP (Ivy) and Canadian GP (Ivy, by three laps from a local man on another Yamaha; indeed all places down to sixth were filled by Yamahas in this event).

Hailwood had a new 350 for 1967 — a 297cc six based on the 250 and developing about 65 bhp at 17,000 rpm. With Redman out of racing, Honda heaped responsibility on to Hailwood for the season, giving him the sort of load that had been almost too much for Redman a couple of years earlier. Hailwood was expected to ride in 250, 350 and 500 races

in the classics; only the company's decision to drop out of 50/125 competition saved him from extra work in those classes! But he was well paid; he was a magnificent rider; and of course he liked winning. So determined was he, so gifted, that he won all his races in the Isle of Man that year, with record laps in each.

The sixes were superb: fast, though probably not as fast as the Yamahas; reliable and, at least in comparison with the two-strokes, reasonable handlers. Certainly Hailwood had no difficulty in controlling the six. He finished the season with five consecutive wins in the 350 class, well ahead of Agostini and the MV. But Yamaha in the 250cc class, with Read, was a much tougher proposition. Hailwood finished only a few points ahead.

The 250 and 350 classes were not his main worry; it was the 500 that depressed him. Giving around 100 bhp, as before, with extra tubes welded into the frame at suspect spots, the RC180 remained too heavy, too high and too uncertain in its handling. Still, he won five grands prix on it, as many as Agostini managed on the MV-3. It was at Monza, at the Italian GP, that the decider was run. When seventeen seconds ahead of Agostini, Hailwood was sidelined with gearbox

Teammates in name if not in spirit, Bill Ivy leads Phil Read, both on RD05A water-cooled Yamaha vee-fours, in the 1967 Ulster GP.

Honda's 297cc six was a bored and stroked 250 and developed 60 bhp. It was Hailwood's favourite Honda.

Mike Hailwood demonstrates a neat cross-legs style while taking his Honda 6 away at Brands Hatch in 1968. On the 297 he won the 1967 350cc world championship.

trouble. For the second year he had failed to bring Honda a 500 title.

At season-end Honda had 250 and 350 titles, thanks to Hailwood's efforts. Naturally, perhaps, the management in Japan found this something of a letdown after their 1966 bag. There was talk of Honda withdrawing from racing. Suzuki had already done so. They had declared they had accomplished all that they had set out to do — which, judging from the unhappy story of the 250 square four, was patently untrue. Now, the official statement continued, was a fitting time to withdraw, before the works racers soared

even further into a never-never land of sophisticated engineering that could have no relevance to the production machines which earned the factory its day to day income. So much for the thesis 'racing improves the breed' which, if it has any substance, should remain valid in all situations. Suzuki were saying that racing cost a lot of money. A stage had been reached where money put into racing could not be justified by evidence of bigger sales. With Honda the situation was similar though not the same.

Honda, as market leaders, were more prestige-conscious than Suzuki and

Yamaha and had never shown themselves particularly concerned by the amount of money involved in running a race team; above all, they were irked by their failure to take the 500 title from Count Agusta of Italy. In the end, withdrawing was largely due, as was claimed at the time, to the effort being channelled into the racing car project, which was intended to achieve for the growing car side of the Honda enterprise what Redman, Hailwood and Co. had helped to do over six years for the motorcycles. That; and the feeling that without a vast expenditure of more time, money and effort Honda might be left at the end of another year with even fewer titles than in 1967. The Yamaha V-4 was clearly superior to Honda among the 250s — possibly 15 bhp superior; so too were the two-stroke 125s; and the 50 class had turned into a Suzuki benefit. And who was to say what MV Agusta might do with their 500? Only in the 350cc class could Honda look forward with some confidence. Without a full-blooded commitment there could be no future for Honda in world racing. Thus it was, in the winter of 67/68, that the announcement was made that Honda would go racing no more.

Indefatigable writer Jeff Clew has revealed that it was Suzuki's practice to destroy their racing machines when they became outdated. With the end of their 1960s involvement there must have been more than a few priceless artefacts to be wheeled along to the incinerator — or whatever was favoured for funeral rites at Hamamatsu. Possibly Honda were more sentimental — unlikely — or were not concerned that their racers should fall into 'wrong' hands. One or two, at least, have survived. A 297cc six formed a snappy backdrop to sundry generators and mopeds in the entrance hall of Honda's UK headquarters. An RC180 500-4, presented to Hailwood in 1968 and left to rust by the maestro when he departed on his car-racing round, was rescued, restored and sent back to Hailwood to display in his Birmingham showrooms.

What about Yamaha? For them, racing without the others would be shadow boxing, expensive and inconclusive. Finally, however, they decided to continue backing grands prix in 1968. It was a year more memorable for personalities than for technicalities (though the 125 V-4 was known to have been developed to 325 bhp/litre, the highest specific output recorded in the class). The personalities belonged to team-mates Read and Ivy;

'mates' perhaps being a misrepresentation. Throughout the season they were at loggerheads, maintaining a thunderous silence when talk was called for, shouting at each other in moments when they were not racing or, more often, dashing off impassioned letters to the motorcycle press. Not the easiest man to rub along with, Read had come to resent in 1967 what he considered to be preferential treatment for Ivy from their Japanese employer. Ivy, it appeared, was to be supported in the quest for both 125 and 250 championships in 1968, and Read was to take a supporting role. The Japanese did not understand Read. He was not interested in taking second place to anybody. Finally some sort of agreement was hammered out: Read would go after the 125 crown, Ivy the 250.

Within weeks of the season opening, each was poaching on the other's supposed preserve, to the accompaniment of verbal warfare on and off the track. Without Honda it was not competition of the highest order; yet, like a civil war, it was more bitter than anything that had gone before. To the world at large it was very entertaining, filling the void for two-stroke v four-stroke enthusiasts with more titillating

The big Honda of the 1960s was the 500-4, which turned out 90 bhp in its first year and demanded all of Hailwood's skill and courage to control during its rapid but uncertain passage through the GPs of 1966 and '67.

above
Brian Ball on a T20 Suzuki in 1968 on his way to a class win, and second place overall, in a 500-mile event in England.

opposite
In the 1968 250 TT on the Isle of Man: Bill Ivy on a water-cooled Yamaha vee-four on the Whitegates-Ramsey section of the course.

news than had ever come their way when the talk was chiefly of mundane tech-nicalities. Ivy sometimes showed himself the faster man, notably in the Isle of Man where, leading the 125cc race, he lapped at 100.32 mph before observing orders and settling to second place behind Read, and in the 250 event which he won with a record lap at 105.51 mph. However, at the end of the season it was Read who collected not one but both champion-ships; he was comfortably ahead in the 125 and managed the 250 by dint of extra points computed, in their wisdom, by the FIM in the way that had brought Hailwood-Honda the title, and lost it for Read, the previous year.

At the end of the year Yamaha decided to follow Honda in withdrawing support for international racing. In fact, they had little choice. The FIM had decided that racing was to be simplified, with 125 and 250 racers being restricted to two cylinders and six speeds. Effectively this outlawed Yamaha's V-4. They had taken five world championships over the years in 125 and 250cc classes. It was not, perhaps, a record to rival Honda's bag of 18 manufacturer's titles and 16 individual championships; but Honda had been at the business a little longer and had spread their net wider. And, of course, they had had the foresight to employ Mike Hailwood.

Road Racing to the Present

Racing in the decade or more from the late 1960s has been largely a Yamaha, later Yamaha **v** Suzuki, affair. Yamaha had a head start, for the FIM dictate accorded with their production-racer policy, which since 1963 had developed a line of 250 twin cylinder racers, named successively TD1 and TD2 and based, at some considerable remove, on the ordinary YDS2 road machine. The 250 was a parallel twin with port-controlled induction. In roadster form it gave about 22 bhp at 7,500 rpm; with modifications it was turning out 35 bhp at 9,500 early in its racing career and was progressively developed to about 42 bhp, by 1969, when the factory gave up official involvment in racing. Impressive though this power output from a basically standard layout may appear, it pales by comparison with the V-4's final statistics of 75 bhp/14,000.

By 1971 the TD2 had become the TDB2, with the piston-port engine having a change in dimensions from 66 x 50mm to 54 x 54mm. Soon the TD series was joined by a 350cc version, the TR2, based on the roadster R3 but with 61 x 59.5mm engine giving 52-54 bhp at 9,000 rpm. The frame followed the pattern of the TD2 and the gearbox had five speeds. At first these racers were reserved for the USA, mainly Daytona, but when released for European competition were an instant success — understandably, as they were the only production racing machines readily available — and began their winning ways, which continue to the present. In 1969 Rodney Gould (later partner with Mike Hailwood in a motorcycle business in Birmingham) rode in several world championship events. Though he received some support from Yamaha HQ at Amsterdam, he did not achieve any worthwhile results. However his team mate, Kent Andersson, managed second place by season-end to Kel Carruthers, of Australia, on a Benelli four.

It was in 1969 that Kawasaki, a comparatively new name to GP racing, made their mark, with a 125cc championship win by the Englishman Dave Simmonds, who was riding in the world championship for the third time, though on this occasion with rather less support from Kawasaki than he had enjoyed in 1967 and '68. His water-cooled 30 bhp twin was extremely fast — fast enough to give him a Belgian GP win at no less than 106 mph — and he took six successive

rounds in the contest, with the title safely his as early as the East German Grand Prix at Sachsenring.

The following year, 1970, Gould made no mistake and won the 250cc world title on his TD2, with Carruthers as runner up. The latter also finished as number two in the 350 class, where his TR2 was clearly almost as fast as Agostini's MV-3. In the 125 class Dieter Braun of Germany, who had trailed Simmonds during the previous year, secured a win for Suzuki.

In 1971 Phil Read, long released from his factory contract with Yamaha but still a rider to be respected (not least by his own reckoning), contested the 250cc class on his Helmut Fath-prepared Yamaha and won it by the fourth round. It was his fifth world championship win, and the first by a 'privateer'. Yamaha continued winning during the next two years with a new star,

left
Production race, Braddan
Bridge, Isle of Man, 4 June
1974. Unmodified road
machines, especially when
they are very fast, are not the
easiest bikes in the world to
control at racing speeds on a
narrow, undulating course.
This fact was made
abundantly clear in the case of
number 25, a 903cc Kawasaki
ridden by Derek Loan.
However, he controlled the
twitching big four well
enough to finish 11th and win
a well-deserved (1,000cc class)
bronze replica.

below
Talk in the paddock. Two
world champions, 10 years
between them. Sheene and
Hailwood, with Sheene's
father in the middle.

Jarno Saarinen of Finland who, after
climbing into the top three in the 1971 250
and 350cc categories, went on to take the
250 class in 1972 on one of the newly
introduced water-cooled models and then
died in a horrific multiple pile-up at
Monza in 1973. It was in 1973 that the twin
in general production became the water-
cooled TZ series. Both 250 and 350 had a
common stroke of 54mm, the crankcase
being identical for both units; bore size for
the 250 continued at 54mm, while for the
larger capacity engine it was 64mm, with
chrome-plating for the bore.

Five-port configuration was carried
forward into the water-cooled versions, in
which the crankshaft was supported on
four bearings, with the righthand end
having a takeoff point for the water pump.
Ignition was electronic, powered from a
generator at the lefthand end of the

93

above
Jack Findlay on a 500 Suzuki
in the process of winning his
first championship grand
prix, the Ulster of 1971, at
95.01 mph.

crankcase. Front forks were telescopic,
with pivoted-fork rear suspension on a full
cradle frame. Brakes were drums, soon
changed to discs.

Saarinen, Gould and Andersson were
factory-backed riders in 1973. With
£30,000 a year from his masters, Saarinen
was engaged not merely to repeat his
250cc class win (and possibly improve his
350 second place . . .) but to topple long-
time 500 champion Agostini from his
perch. For this purpose Yamaha built a
500. It was arranged by the expedient —
simply stated, in reality more than a little
complicated — of setting up two TZ 250
engines across the frame on a new,
common crankcase. Before the grand prix
season was under way Saarinen won the
Daytona 200 in America with his 350
Yamaha, taking a record prize purse of
around £8,000. The first outing for the 500
four was at the French GP at Paul Ricard
in southern France, where Agostini and
Read, on another MV-3, were outpaced
by the new machine. Within weeks
Saarinen scored another win, at the
Austrian GP, and the MV Agusta people
— Agostini, principally — knew that their
durable four-stroke could not hold the
80 bhp Yamaha. It was true too that
Saarinen brought enormous single-
mindedness, courage and skill to his
racing that made him difficult to beat in

right
A difficult man to beat . . . the
Finn, Jarno Saarinen, on a 350
Yamaha in 1972 when he
finished second in the world
championship.

Australian Jack Findlay, who in the mid-1970s, at an advanced (for racing) age of 35-plus, was taking a Suzuki through the world championship rounds. Here he is in the Formula 750 TT. He took the title in 1975, with Barry Sheene as runner-up.

any situation; it was he, above all others, who exemplified the 'climb over the bike' technique for negotiating corners at the quickest rate. Had they been on equal machines it is doubful whether Agostini would have been able to hold Saarinen. Soon after the Finn was killed Agostini did the sensible, if unexpected thing of sounding out Yamaha about riding for them. He had won the 350 crown, for the sixth time, for MV. But the 500cc championship which had been his for seven years, since Hailwood left MV to ride for Honda, had gone to Read on the other MV.

It was in 1973 too that Suzuki showed an official interest in racing, five years after retiring from the grands prix, with a 500cc port-controlled water-cooled twin based on the road-going Cobra. On this machine Jack Findlay won the Isle of Man Senior TT and finished fifth in the world championship.

When Agostini signed to ride for Yamaha in 1974, it was after much cloak-and-dagger work by Rodney Gould, who had given up racing Yamahas in 1973 to work for the firm as their PR man in Amsterdam. Agostini had spent all his racing career on four-strokes yet he took to the 500 Yamaha seemingly without difficulty. The engine characteristics of the two machines were sharply different, with the Yamaha developing most power at around 9,000, in contrast with the MV which had easy power available from as low as 6,000 rpm.

Agostini's first race was the Daytona 200 in March 1974. Yamaha had built a 700 which, like the 500, was a four derived from doubling up twin-cylinder units. And like the 500, it was equipped with new rear suspension patterned on the monoshock system developed during the previous year for Hakan Anderssen's Yamaha 250 moto-cross bike. Power of the 500 in 1974 was given as 95 bhp, with the 350 twin at 56/58 and the 250 up to 45. Agostini ran away with the race, confounding critics who thought of him as star rider only as long as he was on an unbeatable bike — Count Agusta's MV — and something of an also-ran when equipped with more ordinary, or at least more widely shared, machinery. He won the race; and lost the Yamaha. AMA regulations governing motorcycle sport in North America said that a winning machine in any of the big internationals could be claimed, for no more than £2,500, by any outsider putting in a bid within half an hour of the end of the event. Similar situations had arisen in the past, notably after John Cooper's famous win in 1971 at the Ontario race in California, An effective system of counter bidding by the winning rider and/or friends and factory associates had defeated this ludicrous regulation over the years. With the Japanese involved there was much more at stake, not merely in financial terms (because of the enormous costs of producing a Daytona winner in the Japanese style) but in the affront to their code of secrecy. In 1974 the counter-bidding ploy stumbled, and the 'auctioneer' knocked down Agostini's multi-thousand-pound racer to Patrick Pons, a young French rider. Later Pons realized he had far more to lose by incurring the enmity of Yamaha, the biggest force in international racing, than he had to gain from ownership of their up-to-the-minute 700. He returned the bike

and, presumably, collected his money. A few years later Pons, riding a factory F750 Yamaha, was to become France's first world champion.

Though most of 1970s road-racing — particularly for 'private' (or near) riders — was dominated by Yamaha, some part was played by Suzuki following their return to the grand prix scene. Moto-cross had taken most of Suzuki's energies after their withdrawal in 1967. It was only in 1973 that (as noted earlier) the factory ventured into road-racing again, with a developed water-cooled Cobra twin for the Australian, Findlay. Fast it was; but of course it could not hold the MVs or, often, various overbored 350 Yamahas. A more suitable case for treatment in Suzuki's estimation was a somewhat newer road machine, the 750cc water-cooled three. which in standard condition gave approximately 67 bhp at 6,500 rpm. The R and D department at Hamamatsu evolved a very competitive 110 bhp 170 mph racer from the humble GT750.

The porting, inlet and exhaust, was modified to extend the opening period, and the combustion chambers were re-shaped for squish effect and an increase in effective compression ratio. With the fitting of electronic ignition it was possible for the crankshaft, normally carrying alternator and contact breaker, to be shortened, reducing overall width of the engine. The clutch housing was opened for cooling, and ratios for the five-speed gearbox were set closer. Larger-bore Mikuni carburettors and

Three-cylinder 750 Suzuki at speed in 1975: the rider is Stan Woods.

97

left
Crowded start grid. Mick
Grant, Kawasaki, is No 10.

racing pistons were fitted, and the
exhaust system was tuned to incorporate
three expansion chambers. Useable
power for the TR750, as it was called,
extended from 5,500 to 8,500 rpm.

This machine was designed with the
new F750 class in mind, for homologated,
if modified, production models. Ridden
by Barry Sheene, it proved very successful. By the end of 1973 the Sheene-
Suzuki combination had carried off both
the European F750 championship and the
UK-only Superbike series. It was this
class of racing that chiefly interested the
fourth Japanese factory, Kawasaki, which
had taken scant interest in grand prix
racing after the death of the one-time
125cc champion, Dave Simmonds, in 1972.

Kawasaki were anxious to make a
racing impression in the USA (Suzuki too
had angled their effort at the States, and
Sheene had received his conquering 750
from the company's USA subsidiary).
Racing successes would boost sales of the
road machine. In 1971 Kawasaki introduced a larger version of the Mach 3, in
the form of the 750cc H2, shortly to be
modified to racing trim, when it was
raced with great flair by the Canadian,

left
Despite the successes of Kork
Ballington and Mick Grant,
many race followers would
unhesitatingly claim French
Canadian Yvon Duhamel as
the most striking of all
Kawasaki's riders over the
years. Certainly he fell off the
'Green Meanies' more often
than most other contracted
riders. Only an inch over 5
feet, Duhamel had a riding
style that seemed often a
personal war with the
machine he was riding. He
first rode for Kawasaki in 1971
and was still on the track in
1977, at the age of 37, when he
won on a GP 250 at Mosport
Park in Canada.

opposite
Double world champion
Barry Sheene on his Suzuki at
Donington Park.

Yvon Duhamel, at the Talladega 200 and by Paul Smart, brother-in-law to Barry Sheene, at Ontario, California. The H2R was painted bright lime green and Duhamel and company were decked out in matching leathers, which helped to make the racing Kawasakis a distinctive entity on the track (and gave them their label 'Green Meanies'). In 1973 Kawasaki in the USA, with Duhamel, Art Baumann and Gary Nixon as team riders, set their sights again towards Daytona but this time they were drawn against, among others, the extraordinary Saarinen on his 350 Yamaha. The Green Meanies suffered misfortunes with crashes and engine failure when highly placed. As they fell out Saarinen moved up, finally to take the lead.

But apart from Daytona the big Kawasakis had a satisfactory year in America, winning five out of nine important national events. They were less successful in Europe; Sheene and his Suzuki were too good. In 1974 Kawasaki reduced their racing activities. Only Duhamel was kept on as a works-supported rider, with one mechanic, Randy Hall. A year earlier the team had numbered up to five regular riders, with a road manager, Hall as mechanic, and several others. The cutback reflected a downturn in the USA retail market which possibly affected Kawasaki rather more than other Japanese concerns, because of the self-supporting policy that followed the opening of Kawasaki's plant at Lincoln, Nebraska, the first Japanese motorcycle-manufacturing centre in the USA. The H2R was not developed over the winter of 73/74, which would have been normal practice. A team member complained, 'Aren't our motor cycles horrible? We have to use all the old stuff from last year. Look at the equipment on the Yamaha team. It's embarrassing to be in the garage next to them'. Even the balanced Hall was moved to say, 'Our bike's a little faster than last year . . . everyone else is a lot faster'.

In grand prix racing Agostini made a fine showing during two years (1974 and 1975) as a factory rider for Yamaha. He could not, however, take the 500cc championship from Read, his erstwhile team mate, during the first year. The Yamaha was faster but engine problems sidelined Agostini on one or two crucial occasions. At the Italian Grand Prix — particularly important to Agostini, who had alienated much of his countrymen's support by his 'defection' to a Japanese firm — a miscalculation about fuel consumption in what turned out to be a very

fast race brought the thirsty Yamaha to a halt, while in the lead, at the start of the last lap. But on his TZ twin, Agostini collected yet another 350cc championship.

This was the year when Barry Sheene, in intervals in his F750 programme, took to the European grand prix circuits with a more advanced Suzuki than the water-cooled Cobra. It was a water-cooled four, patterned on the 250 square-four of the mid-1960s which had caused Frank Perris and others so much heartache. Disc inlet valves were fitted in the 56 x 50.5mm unit, and the crankshafts, running in needle and ball bearings, were geared to an intermediate shaft and thence to the dry clutch and six-speed gearbox. Water impeller and ignition were driven on the left. Lubrication was by petroil, ignition was electronic, and the exhaust system had expansion boxes under the engine for the two front cylinders, and straight pipes for the other cylinders, arranged

above
Finally a world champion 15 times over, Giacomo Agostini took to Yamaha two-strokes, after years of success with well-mannered MV Agusta multis, without apparent difficulty. In 1975 he won Yamaha's first 500cc world championship.

left
A countryman of Jarno Saarinen, Tepi Lansivuori of Finland remained in the Yamaha camp when Agostini took over as leader following Saarinen's death at Monza in 1973. But not for long: he resented Agostini's favoured position, and moved to Suzuki.

overleaf
American veteran Gary Nixon on a Kawasaki.

above
Twice a world champion on Suzukis, Barry Sheene has been the outstanding British GP rider from the early 1970s. In 1980 he moved to Yamahas, first on privately maintained machines but latterly as leader of the official Mitsui Yamaha team.

right
Johnny Cecotto arrived on the road-race scene in a big way in 1975 as a 19-year-old from Caracas, Venezuela. He rode for the Venezuelan Yamaha importer, Andres Ippolito, and was too good for Giacomo Agostini in the 1975 350cc world championship.

just under the seat. With 34mm carburettors, the engine gave 90 bhp at 10,500 rpm. Despite the 500's erratic handling Sheene and Findlay had a reasonable season, with a second place for Sheene at the French GP, and fourth and fifth places shared with the Australian which eventually hoisted Findlay to the higher position in the championship, at fifth.

In 1975 Agostini took Read very seriously, and won the 500cc world championship for Yamaha; it was Yamaha's first in the top class of GP racing. The 350cc title went to a Venezuelan, Johnny Cecotto, only 19, who had arrived in international racing with a startling third place in the Daytona 200, following this with 250 and 350 class wins at the French Grand Prix at Paul Ricard. Agostini on a 350 seemed unable to cope with Cecotto; he won only one 350 round, at the Spanish GP, though his succession of runner-up places was enough to make him number two by the

end of the season. Read won two of the 500cc rounds and had several seconds but was outpaced by Yamaha-mounted Agostini. Towards the end of the season it was Sheene, with his RG500 Suzuki, who was getting the Italian's attention.

Sheene was riding under the auspices of the Heron Suzuki GB organization. His bike had been developed in a number of ways and at the beginning of the year the Suzuki people had high hopes of a championship win. These were shattered when Sheene crashed at about 170 mph in February while practising for the Daytona 200,, receiving injuries that kept him out of racing for almost two months. When he came back, limping and severely weakened, he was apparently undiminished in spirit, style or skill and ended the season with two grands prix wins, in Holland and Sweden. His teammate was Tepi Lansivuori, who had been in the Yamaha line-up with his countryman, Saarinen, and later with Agostini with whom he had been on fairly distant terms on account of his — Lansivuori's — much smaller race fee from Yamaha. When a Japanese rider, Kanaya, was drafted into the team, Lansivuori took it as a personal affront and went off to Suzuki. (Yamaha's choice of Kanaya was rather more than blood calling to blood, for he rode very well through the season and even won a 500 round, in Austria.) Sheene suffered various mechanical afflictions in 1975 in addition to the physical shortcomings he had to live with following the Daytona crash. There were numerous troubles with gearbox and primary drive, though when the Suzuki was going it was very competitive, its disc-valve engine fully a match for the piston-port Yamaha. Power had been raised to about 105 bhp at 11,500 by dint of port and compression ratio changes and shortening the stroke to make the engine 'square' at 54 x 54mm, the classic dimensions established years before by Kaaden, the MZ designer. Weight was saved and the fairing contours were improved, so that top speed was upped to an estimated 180 mph.

For 1976 Sheene had John Newbold, Percy Tait and John Williams as back-up members of the Suzuki race team, by then enjoying sponsorship from Texaco Oil and Forward Trust Finance. Yamaha had chosen to withdraw official support for grand prix racing; instead, their name was carried by the 1975 350cc champion, Cecotto, who took 350s and 500s around the world under the banner of his home team, Venemotos. But Johnny Cecotto was no substitute for Agostini. A brilliant rider, he lacked the dedicated approach of the Italian — or for that matter of Sheene — and his record through the season was, by Yamaha standards, lamentable.

Highly paid (his earnings were put at £200,000 a year) and lionized in his own country, he rapidly acquired much of the big-star temperament and found the routine involved in getting to meetings, liaising with technicians and officials — generally, the mundane, day-to-day part of a road-racer's life — too irksome. More to his liking was something like the Daytona 200 where, in 1976, he and several others mounted on the latest 750cc OW31s completely outpaced Sheene on the 750-3 Suzuki. In the grands prix, however, Sheene proved that the RG500 was more than able to hold the Yamaha — and that he was master of the mercurial Cecotto. He won the first four events in the series and at the Swedish GP, in July, took the title. The power of full works support was made clear when Suzuki, unplaced in 1975, filled the first three

above
American Steve Baker
overtakes John Newbold,
both on Yamahas.

places in the world championship, with Lansivuori and Pat Hennen of the USA behind Sheene.

Yamaha had concentrated on 750s in 1976. Some years earlier, while developing the 500-4 for Saarinen, the Yamaha administration had realized the publicity value to be gained from racing seven-fifties. Formula 750, at the time run only as a European championship, would eventually be given 'world' status; and in any case this was the era of the superbike, as pioneered by Honda and Suzuki. Yamaha's first superbike, the 700, was in effect a double-up of TZ350 twins and was said to produce at least 90 bhp at 10,000 rpm. As recounted, it was on one of these that Agostini won the Daytona 200 in 1974. He was not the only man at Daytona on a big Yamaha; the factory had rushed through a big batch for homologation. In Europe they were available to selected riders at around £4,000 each, and were clearly too powerful for the TR750 Suzuki and the Kawasaki, operating on three cylinders and feeling the pinch. Development of the 700 Yamaha continued, with the engine being enlarged to a full 750cc. As a works racer it produced 110 bhp, as a production series around 100; the works model had monoshock suspension, the others ordinary pivoted fork. By late 1975 Yamaha had taken the 750 a stage further, with a general lightening of the chassis through extensive use of light-alloy and magnesium, even titanium.

left
Agostini in England at the so-called Race of the Year.

107

Japanese riders have a mixed record in road-racing. With some few exceptions, they have not proved a match for the best of the Europeans. One of the exceptions is Takazumi Katayama, seen here on his 250 Yamaha at Quarter Bridge in the 1976 Isle of Man 250 TT. The following year he rode extremely well, mainly on a newly developed Yamaha triple, to become 350cc world champion — the first, and to date only, Japanese to take a road-race world title.

Early big Yamaha: a 700, in effect two 350s coupled to produce 90-100 bhp.

Bore and stroke remained at 66.4 x 54mm, for 748cc capacity, and the gearbox was a six-speeder with variation in ratio available on several gears, and braking was by three 12in diameter discs. The engine had reed valves and developed close to 120 bhp. The expansion-chamber exhaust system had two boxes on the right, two on the left, with one of the latter crossed over to sweep back to the right for optimum effect. OW31 was the name for these 750s, and initially only a handful were manufactured for use in the 1976 Daytona classic where, as noted

earlier, they were too fast for Sheene. The power and reliability of the design was shown in 1977 when the American, Steve Baker, used a works machine reputed to give 130 bhp to take the first world F750 title after a succession of scintillating rides on, for him, mainly unfamiliar circuits in Europe as well as in California.

Agostini had faded from the scene in 1976. Returning to grands prix after a year's sabbatical, Yamaha chose Baker and Cecotto. The YZR500 had been developed for 1977 in line with the OW31 and for the new season was said to

produce around 100 bhp, which was rather less than the output of the leading Suzukis. The limitations of the Yamaha's symmetrical port timing were becoming apparent, with the power band being forced ever higher and narrower as peak revs approached 11,500. Sheene was on top form through 1977, while Baker and Cecotto struck trouble, in some measure of their own making, involving well-publicized quarrels with various camp followers, which must have outraged the formal boardroom of Yamaha. Cecotto crashed in the Austrian GP, breaking an

arm, and was not able to ride much in the second half of the year. Baker too fell on several occasions, and was reduced to trailing Sheene in the 500 classification.

The 350 class had seen some technical innovation with a three-cylinder developed by European Yamaha in an effort to dislodge Italian Aermacchi (using the parent factory's name, Harley-Davidson) from the head of the championship table. Walter Villa, the Harley-Davidson rider, had taken world 250 titles in 1974 and 1975, and for 1976 the factory had given him a 350 as well, which enabled

Alex George, later to win — and crash spectacularly — while riding for Honda in the Isle of Man in 1979, treats Ballaugh Bridge in the traditional, tyre-saving fashion favoured by the fast men. He is riding a Yamaha and the event is the Junior TT of 1975.

him to run away with two titles, 250 and 350.

Apparently resigned to a drubbing among the 250s, Yamaha were loath to accept second place in the 350cc class. The TZ350, the definitive 'privateer' racer, was not considered a fair match for the Italian machines: too few cylinders and too little power were handicaps that even Agostini, riding in a 'semi-private' capacity, and Takazumi Katayama would be unable to surmount. Hence the decision to make a three. The engine used mainly standard components from the 250 twin. The crankcase was basically a TZ250 twin's with an extra block welded to the lefthand side in a manner devised by the Swiss, Rudi Kurth, who had previously made a three in 500cc capacity. The three was offset 10mm in the old 350 chassis for chain alignment. The crankshaft, produced by a German firm, Hoeckle, carried standard TZ connecting rods and big-end bearings. With 54mm bore and 50.8mm

stroke, the engine was a full 350cc. TZ350 six-speed transmission and clutch were used, and water impeller-pressurized coolant was carried in a large (TZ750) radiator. Power of the triple worked out to approximately 80 bhp, almost 10 bhp up on a TZ350 twin.

The added weight of the three over a twin was a disadvantage, though handling, once the unit was shifted from the standard 350 chassis into a frame made by the Dutch specialist Nico Bakker, was considered to be 'very reasonable'. Both Agostini and Katayama rode triples in the West German GP at Hockenheim and raced away from the opposition to score a one-two victory, with Katayama ahead. Finally Katayama took the championship.

The 250cc class was enlivened, in 1977, by the presence of the Kawasaki in-line (tandem) water-cooled twin designed by Nagato Sato two years earlier. With disc valves and 34mm Mikuni carburettors, it produced 55bhp at 11,500 rpm. The pistons

were arranged at first in out-of-phase sequence — when vibration proved intolerable — and later in step, which diminished the vibes though at a slight power cost. The engine had an aluminium block and head and was mounted in a duplex frame with rear suspension incorporating a bell crank, or rocker arm, to transfer pivoted-fork movement to a spring-and-damper strut positioned vertically behind the gearbox. Later the 250 was enlarged by overboring, to compete as a 350.

It was on the 250 Kawasaki that Mick Grant achieved his first grand prix win. He joined the Kawasaki team of Kyochara and Barry Ditchburn at the French GP in May, and at one stage appeared a likely winner, until transmission failure brought him to a halt. At the Dutch TT at Assen, with plenty of bends and few straights, the Kawasaki's remarkable acceleration and angle of lean, made possible by the narrowness of the tandem twin, offset the

above
Mick Grant, a redoubtable performer in the Isle of Man on Kawasaki.

opposite
Patrick Pons, who won France's first world championship, the F750, in 1978.

right
Suzuki 750-3. This durable water-cooled racer, based on the roadgoing GT750, was usually too fast for the big Kawasaki in both Europe and the USA.

below
Steve Parrish, sometime car salesman and a protegé of Barry Sheene, seen during his 1977 season with Texaco Heron Team Suzuki. That year he finished fourth in the world championship and — fourth places apparently not being good enough for his employers — was dropped from the team. In 1981 he returned to works sponsorship and further association with Sheene as a member of the Mitsui Yamaha race team.

speed superiority of the leading Yamahas and Harley-Davidsons.

Road-racing is nothing if not a reflection of rider temperament. Examples to prove the point could be unlimited . . . but consider Redman as the cool, calculating tactician; and Saarinen riding always to the very limit of his ability, and beyond most other riders', to achieve a sort of suicidal consistency; and Cecotto as an amalgam of South American and Italian volatility, which meant an inspired performance on some occasions and no better than a miserable flop on others. Grant, the not-so-stolid Yorkshireman, found motivation for his Dutch effort in what he saw as discriminatory treatment which had him racing the 250 late in the season, after Ditchburn and Kyochara had been able to make their mark (which they had done fairly effectively with near-wins in the W. German Grand Prix, the Italian at Imola and the Spanish.) Grant was considered — dismissed, in his view —

as a 750 specialist more at home on long, undulating 'natural' courses like the Isle of Man, racing against the clock, than competing against top-class talent in the hot-house atmosphere of the grands prix. Determined to show he was more than this, he set a cracking pace in the Dutch. When rain came early he ignored the slides provoked by the fair-weather 'slick' tyres on the Kawasaki and finally won by 11 seconds from the Harley-Davidson rider Uncini. Grant found his win particularly satisfying — and galling. While it confirmed his feeling that he could have done a great deal with the 250 in the first weeks of the championship, he thought there was little hope of managing a class win with only five rounds remaining. He was right. Though he scored another win, in Sweden, and a second place in Finland, the few rides he had through the season, and complete failures in other rounds in which he started, kept him out of the final count.

George O'Dell, sidecar world champion in 1977, and later to die in a well-publicized house fire incident in March, 1981.

113

Kenny Roberts, three times a world champion, has had only one win at Daytona, in 1973.

The Japanese took over another branch of grand prix racing in 1977, apparently in perpetuity. Yamaha and George O'Dell won the sidecar championship without winning a single event of the series. If O'Dell had not managed it (to give Britain its first world sidecar title in 24 years), Yamaha still would have triumphed, for behind the Englishman was Rolf Biland of Switzerland, on another Yamaha. O'Dell won despite a string of setbacks that apparently only spurred him to greater effort. For much of the season he was without his number-one outfit, a Seymaz Yamaha equipped with hub-centre steering and car-style rear suspension; he rolled it six times in a non-championship event at Cadwell Park and thereafter it was only intermittently available, after rebuilds, which meant that O'Dell had to rely on a two-year-old outfit with conventional front fork and,

consequently, a circuit performance inferior in almost every way to that of his 1977-designed bolide.

Sheene won five of the world championship rounds in 1977 and took his second world title, with Pat Hennen, the 23-year-old Californian, and Steve Parrish of the Suzuki GB team to back him up in third and fourth places; Baker on the Yamaha was second. Sheene's Suzuki had been refined a little from 1976, with power up to 118 bhp and air rear springing.

In the years since 1978, hard-working PR men have done their best with the mildly unco-operative Kenny Roberts of the USA, who has won three world 500cc titles (1978, '79 and '80) without compromising his determination to keep public utterances to a maximum of three consecutive sentences on any one occasion. In this perhaps he had done no more than

left
When Kenny Roberts came to
Europe it was as part of a
revitalized team to contest the
annual Transatlantic Trophy
races — USA v Britain — held
over the Easter holiday period
in the UK. One of his Yamaha
teammates was Gene
Romero, a forceful rider well
to the fore in American
racing.

Americans in the late 1970s acquired a reputation for invincibility in Europe. Perhaps they had the essential secret of relaxation? Here Gary Nixon, Romero and Roberts gather their energies for the next outing at Brands Hatch.

TZ500 Yamaha for 1980 fitted with 'power valve'. Power output was in excess of 110 bhp at 10,500 rpm.

follow the tradition of the archetypical American man of action, long familiar to Western film addicts, who has always preferred a meaningful *Yeah!* to long-winded exposition. One result of Roberts' taciturn ways has been that the non-motorcycling public in western Europe has needed most of those years since 1978 to twig that the ebullient Sheene is no longer the champion.

Roberts had raced in Europe before 1978. In 1975 he went to Holland for the grand prix and gave 250 champion Walter Villa something of a fright before overbraking and crashing on a hairpin bend in the late stages of the race, which meant that after climbing back in the saddle he had slipped to third place. In

1977 he turned up in Britain for the annual Easter match races between America and UK, and took four wins from six starts. A few weeks later, at Imola, he won 250 and 750 events. Sheene was one of the British riders outridden by the American at Easter: The euphoria of later successes in the grands prix did not blind him to the threat that would be posed by a full-scale Roberts attack in 1978. Suzuki played their part by developing the square-four to a peak of 122 bhp at 11,000 rpm, and modified the gearbox to facilitate servicing and bring the gearbox sprocket and rear-fork pivot closer, minimizing rear-chain snatch. Front discs were increased in size and the fairing was given two small foils at the front, one each side, for added down-thrust when the front wheel lightened at upwards of 160 mph.

Roberts came to Europe in 1978 with his own sponsors and a brand new bike from Yamaha America. Chef d'équipe was one-time world champion Kel Carruthers who had early recognized Roberts' quality. The équipe apart from Roberts numbered usually no more than one: mechanic Nobby Clarke, whose impeccable references included service in the 1960s with Hailwood and other Honda heroes. The 500 Yamaha was, as were previous Yamahas, an in-line four with symmetrical port timing. That it was a match for the disc-valve Suzuki was due to intensive research into the inertia and pulsation effects of gas flow as influenced by the exhaust and scavenging systems.

To achieve even higher revs the Yamaha engineers had constantly to revise the

South African Kork Ballington on the ultra-slim, water-cooled 250 GP Kawasaki. He won both 250 and 350cc world championships in 1978 and 1979.

port timing, with the height of the exhaust port increasing with every few hundred revs gained. This was all very well, apart from an unwelcome narrowing of the power band, which could mean useful power being restricted to no more that 1,000 revs at the top of the scale, and the deleterious effect this had at low road speeds when the rider was required to juggle rpm in a high, critical band. Yamaha had the idea of introducing a variable-height control for the exhaust port which would enable the port to be lowered at medium rpm, for a broader power band. They called it the Power Valve. It was a shutter at the top of the port and it was of course controlled by engine speed, as indicated by the tachometer. As revs dropped the shutter moved, effectively lowering the height of the port and deflecting into the cylinder gas which would otherwise escape through the port. The effect was probably minimal, and setting up the shutter was a delicate business, but the device was thought to be necessary to offset the flexibility achieved in later development of the rival Suzuki design.

Roberts was not a North American replacement for Steve Baker, who found himself out in the cold after taking second place to Sheene in the 1977 contest. An experienced commentator has said, 'It is not generally known that Yamaha can be infinitely kind and helpful to men who have brought them recognition and success'. Equally, it appeared they could be infinitely slow in renewing the contracts of men who brought them only partial success. The official factory-

backed rider was Cecotto, with several bikes and mechanics. However, partway through the season Yamaha, disappointed with Cecotto's showing, transferred support to Roberts.

Kork Ballington of South Africa took the tandem Kawasakis to well-deserved success in 250 and 350cc world championships. Proof that these wins were not dependent solely on the skill of the South African was shown by the German Anton Mang's 250cc championship win for Kawasaki in 1980 and the outstanding rides of Gregg Hansford of Australia (twice a 250cc-class runner-up).

In road-racing circles during the last two years of the 1970s more attention was paid to what turned out to be a dismal

Two-stroke tandem twin Kawasaki in 250 form as raced by Ballington to his first world championship, in 1978. The front crankshaft has the generator for the electronic ignition on its righthand end; at the rear, driven by the other crankshaft, are the clutch, water impeller and rev-counter takeoff.

overleaf
Biland's Yamaha-powered bolide at the French Grand Prix, 1979.

World champions in 250 and 350cc classes, Kawasaki have set their sights on the 500 category, with Kork Ballington riding a four. Pictured is the 1980 development.

opposite
Thoughtful-looking — no, let's say dejected-looking — Mick Grant on the first NR500 Honda to arrive in the UK, for the British Grand Prix at Silverstone in 1979. The vee-four four-stroke revved to 18,000 but was too slow to challenge two-stroke opposition.

flop than to the winning exploits of Yamaha and Suzuki, Roberts and Sheene, and the others. The also-ran . . . in fact it ran hardly at all . . . was the work of Honda, who had announced, in a low-key way, that they would be back in grand-prix racing in 1978.

Though the world's number one producer had turned away from an exclusively four-stroke programme, having made successful forays into moto-cross with advanced two-strokes, the name of Honda was apparently still indissolubly linked with the four-stroke. Tokyo decided it would be seen as a betrayal of traditional marketing philosophy to make a two-stroke for road-racing. (It would be unwise to underestimate the weight of Soichiro Honda's views in this matter. Though retired from day-to-day involvement in company affairs, Honda continued to exercise power in the boardroom. It is consistent with his determined personality that the decision was made to follow the hazardous four-stroke path in building a rest-of-Japan beater.)

Drawing on their experience of 1960s racing, but employing an entirely new development team, Honda Motor saw only one way of getting near the leading two-strokes; their engine would have to rev faster than any conventional four-stroke of its size had managed before.

Their views on the proper way to achieve this had not altered over the years, and followed classic lines. The stroke would be ultra short, to keep mean piston speed in bounds, and there would be multiple valves for low reciprocating mass and optimum gas intake. The engine would be light, as well as the bike as a whole, which, additionally, would have to be smaller and more 'slippery' than other 500s to help in offsetting the gap that would surely exist between the specific power output of the four-stroke and the best of the rest.

Delays occurred. Originally Honda hoped to have the bike, code named NR500, ready for a race or two in 1978; then it was to be at the Dutch TT in 1979; finally two machines, with mechanics in attendance, arrived at the British Grand Prix at Silverstone for use by Grant and Katayama. The bikes failed, quite miserably. They were slow, handled badly, barely qualified in practice, and retired very early in the grand prix. A similar performance took place at the French GP a few weeks later. Failures though they were, the NR500s provided much of interest for anybody brave enought to ignore the *Keep Out!* cordon thrown up in the paddocks. Very small on 16 inch wheels shod with special British Dunlops, they weighed under 250 lb and

The US rider Randy Mamola wheelies a 120 bhp Suzuki. Now a kingpin in the Suzuki racing team in Europe, he finished second in the 1980 500cc World Championship during his first year away from home.

had the engine installed in an aluminium monocoque supporting trailing-link steering, with spring units carried in front of and parallel with the main telescopics. Rear suspension was by then-commonplace monoshock, with the spring under the seat and heading up towards the steering head. Chain snatch was eliminated by having the rear-fork pivot on the gearbox centre, and additional stiffness was conferred on the aluminium box-section fork by a U loop linking the arms from a few inches forward of the wheel spindle. Later some details of the engine were released. Perhaps it would be more accurate to say that intelligent speculation about the engine were confirmed by Honda's refusal to deny the guesswork. This procedure is baffling to outsiders but passes for a clearcut question-and-answer session on the Oriental race scene. The NR500 had a stroke of 36mm,

would rev to 18,000 rpm (which was about 5,000 short of being useful) and had elongated bores, hence elongated pistons, to accommodate eight tiny valves, two sparking plugs and two carburettors per cylinder. And of course ignition was transistorized. (At about that point speculation tended to peter out.)

The word is that Honda, no faint hearts, are sticking with the NR500. Privileged observers at R and D tell of banks of NR engines costing thousands revving to messy destruction on the dynamometers. When the debris are swept away, almost without a glance, fresh engines are installed and the operation is repeated. Honda silence critics who recall the methodical, penny-pinching ways of pre-Computer Age wizards like Norton's Joe Craig by declaring that they are short of time, not money . . .

above
Works Suzuki rider Graeme
Crosby in 1980. He joined
Mamola in the 1981 Suzuki
team.

left
Latest Yamaha 250 twin road
racer with 'power valve'.

123

Endurance racing

Endurance — long-distance — racing dates back to the beginning of motorcycling. The marathon from Land's End to John O'Groats and the Paris-Bordeaux, in the early years of the century, were endurance races. The 'race' element then was mainly against the clock, in essence another man's time, and the rider's stamina was put to a greater test than were the engines of those tall, spindly bone-shakers.

The Japanese came into the picture rather later, of course — in the late 1960s, when Suzuki's 250 and 500cc two-stroke twins had class wins in the Thruxton 500-Miler event in Hampshire. In 1969 the FIM instituted the Coupe d'Endurance. As makers began to develop their engines in search of more power the capacity limit for the Coupe was raised, in stages, to 1,200cc. The chief appeal of the endurance racer has always been its kinship, sometimes more apparent than real, with the standard motorcycle sold to the public; this, together with sheer size and power, as displayed by the later endurance machines.

Rules for this branch of racing have been decidely looser than those governing grand prix competition. Chassis of outlandish design and complexity have been developed, and later abandoned, in

Poet? Musician? Endurance racer? As this is a book about motorcycles there will be no surprise in the fact that the man pictured is Georges Godier, part of the Godier/ Genoud team that has twice won the Bol d'Or on Kawasakis.

efforts to deploy the enormous power of 1-litre fours, and even sixes, in the peculiar conditions of 24-hour racing on circuits where often a greater premium is placed on sheer speed than on engine torque or flexibility. Before 'Honda Europe' — the title embracing national teams from France, the UK and Germany — took a hand in the game around 1975,

endurance racing had become something of a walkover for specialist entrants who, provided they complied, more or less, with homologation requirements, were free to install Honda, Kawasaki and Suzuki engine units in chassis of their own devising. The Kawasaki 903cc four was extremely popular, with its apparently limitless capacity for absorbing punishment without complaint. Notable Kawasaki exponents were Godier and Genoud who had Sidemm, Kawasaki importers in France, as entrants. The engine, later enlarged to 1,000cc, was developed progressively from 90 bhp to 120 bhp in the period 1975-78. But a problem arose when Honda began to show a special interest in the Coupe d'Endurance. Private entrants and sponsored teams like Godier and Genoud had earlier made use of the ubiquitous CB750-4 unit, gingered up with dealer-supplied race kits, to give an extra 15-20 bhp, and fitted into 'one-off' frames. But this would not do for Mr Honda. The official endurance machines had to be very close to the road bikes. Racing peculiar-looking specials in what was basically a production-machine event would be to admit that the standard Honda was deficient. So while dealer-entered Kawasakis enjoying some small measure of 'official' backing continued in endurance races of the late 1970s to display a technically interesting, and usually very effective, variety in chassis art and development, the factory Hondas were tall and heavy, with apparently standard frames and conventional telescopic front and pivoted-fork rear suspension systems. Advanced thinking had the engine as an integral part of the chassis, and for small-scale constructors there were no marketing considerations against dispensing with the 30-year-old telescopic fork in favour of leading-link or trailing-link or hub-centre arrangements. Rear springing was usually monoshock in some form, with the springing medium housed under the seat or the engine, or atop one of the arms of the rear fork. Usually the fuel tank would be in its accustomed position; occasionally it would be slung *under* the engine, to improve the centre of gravity.

By contrast, factory-backed Hondas appeared painfully ordinary. Where they gained, immeasurably, was in engine development. Within a year of Honda Japan taking a new interest in the class, peak power of the 1,000cc fours soared to 130 bhp, well ahead of the output of the best Kawasakis. Though the specials might be more nimble that the 65 lb heavier Hondas, especially in early days, before Honda lowered frames and strengthened forks, extra engine power and the benefits of factory backing and organization were enough to give Honda an advantage.

The homologation rules specifying the minimum number of machines to be sold before their entry, no matter in what violently modified form, could be accepted for Coupe d'Endurance, were not framed with 'pure' racing machines in mind. But Yamaha, as the major race-machine producer, soon turned out the qualifying number of racers, and in 1974 a TZ700 was running in the Bol d'Or at Le Mans. In 1976 a TZ made fastest lap there but retired half-way through; and two years later an OW31 750 kept ahead for 18 hours. These performances came as something of a shock to Coupe regulars who had assumed that their branch of the sport, with its special demands and conditions, had evolved a special type of machine that could not be challenged. A GP machine, they said, while adequately fast, was not designed or built to last the long hours of these marathons.

In the early 1970s the homologation rules paved the way for endurance race machines evolved from the Kawasaki two-stroke triples and race-kitted Suzuki twins and water-cooled threes. Honda, for some years the only Japanese manufacturer taking a special interest in endurance racing, made tentative approaches in 1971 and followed with an entry in the Spa 24-hour event of 1972, when their team of John and Charlie Williams (not related) was managed by Alf Briggs, for several years a member of Honda GB's staff in London and Nottingham. They used a race-kitted CB750 developing approximately 90 bhp, and they won the race.

The most famous race in the endurance calendar was, and remains, the Bol d'Or, brought back after a lapse of eight years in 1969 to mark the first year of the Coupe, when it was staged at Montlhéry and won by the newly introduced Honda 750. The following two years were success stories for the dying BSA/Triumph empire with their 80 bhp threes; but from 1972 Bol d'Or finishing lists have been dominated by Japanese fours, mainly Kawasaki and Honda. Four riders (two teams) have been kings, their names rolling off English tongues with unaccustomed ease: Godier/Genoud and Léon/Chemarin. (Christian Léon is dead, victim of a bumpy corner and his own right hand on Suzuki's test track in Japan in 1980). Georges Godier, a Frenchman,

mechanic by trade, and Swiss Alain Genoud, one-time barman, turned their hands to chassis construction; and from the mid-1970s they have left riding to other people. On team Kawasakis they were unbeatable in 1974 and 1975 at the Bol d'Or. Their 1,000cc engines were Yoshimura tuned, but not to an extreme degree; instead they relied on overall preparation of the machine, emphasizing light, responsive handling and quickly detachable components to save time at the pits. They changed to Kawasaki, first 903cc, then 1,000cc, from CB750-based Egli-frame specials. In endurance racing, as in drags and sprinting, the Kawasaki four with its roller main bearings and built-in strength proved a favourite for all the top constructors.

top
Honda Britain endurance racer. Changes occur from year to year and, at the same time, bike to bike. This 1976 RCB has brake calipers behind the fork legs and 'right way up' de Carbon rear dampers.

above
Under the fairing. Leon/Chemarin 1977 RCB has calipers ahead of the fork and a lower frame than the '76 version.

above
Fresh from the crate — from
Japan — a 1000cc RSC Honda
is readied for tests at
Donington Park.

Another notable Endurance machine of the early 1970s was the French Japauto, with elaborate fairings and (usually) Honda power units; in 1972 and 1973 Japauto, with Gerard Debrock as chief rider, won the Bol d'Or. When Honda began to take an 'official' interest in Coupe d'Endurance in 1975, six years after the first round of the series in which they had scored with a tuned 750, the competition was much stronger, with the likes of Godier and Genoud and Debrock and even Yvon Duhamel, the Canadian GP rider, on well-developed Kawasakis. Where once a race kit for the Honda 750, to boost power to 75-80 bhp, twin discs for the front wheel and high gearing were enough to give class wins, by the mid-1970s the fours required at least 110 bhp, and custom building, to stand a chance of finishing among the leaders.

Honda's 1975 entries, the only machines to have some measure of official backing from a Japanese factory, had to pay the price, as previously explained, of maintaining close links with the production bikes on public sale. But a condition such as this may be turned on its head; the racer may be the one to be developed first, with a new line of roadsters patterned on the racer's frame and engine to accepted levels of conventional taste. This is the course that Honda followed with their RCB racers, which came along in 1976 with dohc, four-valve cylinder heads and 941cc capacity. On a compression ratio of 11.5:1, these ordinary-looking motorcycles were giving about 115 bhp at

left
Christian Leon, one of Honda
France's invincible duo of
Chemerin/Leon, on the way
to another Bol d'Or victory.

above
Honda France team in 1977. From left: Jean-Claud Chemarin, Michel Rougerie, Christian Leon. The last-named was killed in Japan in 1980.

below
Honda waits for Phil Read. Road-racers make occasional forays into the endurance arena. Read rode with Tony Rutter in the 1977 Bol d'Or.

9,000 rpm. Within 18 months the roadster line, known as FZ 900, had appeared, backed by a 750 version, with Honda undertaking to make race kits available to lift standard power from about 95 bhp to near the level of the original RCB. In addition to dohc and 16 valves, the FZ line carried on the race bikes' FVQ rear dampers, quartz-halogen lighting and Comstar wheels. More important, perhaps, to ordinary, non-racing owners were Honda's improvements in chain construction, called for by frequent

breakdowns in endurance racing. Sealed 'O' ring chains, pioneered on the racing models, made an appearance on the FZ line.

In 1976, '77 and '78 J.-C. Chemarin, first with the Englishman, Alex George, then with Léon, his regular partner, won the Bol d'Or on Hondas. The Kawasaki specials, though ahead in chassis design in the first year, were unable to call on factory backing; their straightforward power unit, with two-valve head, began to fall behind in the horsepower game, to the point where the Honda's built-in handicap of a tall, heavy frame was no longer important. In fact, after 1976 Honda endurance machines profited from lower, lighter frames with a strengthened, triangulated steering head, and engine capacity went up to 997cc, with power at 125 bhp. Chemarin/Léon, riding as usual for Honda France, won their fourth consecutive Coupe d'Endurance title in 1979. Not that they had everything their own way, with a crash, a broken chain and an official protest against them to contend with in the Bol d'Or (which they won).

The Japanese domination in the Coupe is almost total. In 1979 there were only two European contenders, Italian Ducatis, among the first 18 finishers. The Honda has been shown to be almost unbeatable, though the 750 GP Yamaha racer entered by the French Yamaha importer, Sonauto, in the 1978 Bol d'Or run at the Paul Ricard circuit near Marseilles, proved a distinct worry to the four-stroke kings. Faster and lighter than any of the four-strokes, the Yamaha was ridden by Patrick Pons and Christian Sarron. It was indisputably in standard — for GP racing — trim apart from the mandatory array of extra headlights. Its many detractors (later quoted by Yamaha) had plenty to say in advance: 'It will never last. It will go fast for maybe five hours, and then explode'. It was in the lead after five laps, which was not unexpected; but then it was still ahead after 10 hours, and at 16 hours too — until, in the 18th, a crankshaft broke and the two-stroke was out.

In 1980 a 'new' name won the Bol d'Or: Suzuki, with Pierre Samin and Franck Gross as the riders. At last there appears to be some effective challenge, after six years, for Honda in the Coupe d'Endurance, which in 1980 was elevated from European to world championship status, albeit with rather restrictive TT Formula One rules based on the use of more or less standard four-stroke engines with a capacity limit of 1,000cc.

Moto-cross

Moto-cross was popular in Japan soon after World War Two, but at that time showed little of the professionalism that existed in Europe, where events such as the Moto-cross des Nations were an established feature from 1947 for top riders in national teams from France, Holland, Belgium, Sweden and the UK. Early moto-cross machines turned out by the Japanese, as exemplified by the 305 Honda 'scrambler' of 1962, were ludicrously heavy and unwieldy. They would not (had anybody been so misguided as to put the matter to the test) have finished among the first 20 in any middling-calibre event organized by a Sunday-afternoon club.

Modern moto-cross ranks second only to road-racing in prestige, manufacturer-support and popularity among spectators. At some stages in its 55 years moto-cross has attracted more interest than road-racing from the leading factories, mainly because of the generally lower development costs involved in building a cross-country winner, compared with a road-racer. BSA in Britain, for instance, regularly supported international moto-cross when their road-racing commitments were limited to sporadic, semi-'amateur' forays into clubman compet-

ition; and much smaller firms in Europe, such as Swedish Husqvarna and FN in Belgium, have tended to restrict themselves exclusively to moto-cross.

In the early 1960s going into moto-cross in a big way made sense for Suzuki for a number of reasons. First, it would take them into an arena of motorcycle sport free of the dominating presence of Honda and Yamaha, who between them appeared to have the major, attention-drawing classes in road-racing sewn up for the forseeable future. Secondly, moto-cross was likely to make a great impact in the USA — more than road-racing even. Like the other Japanese firms, Suzuki attached paramount importance to Americans and all their ways. Finally — perhaps least important — moto-cross would not be as expensive as road-racing. Their rider Matsujisa Kosimo took a 250 Suzuki to Europe for a few events in the 1964, '65 and '66 seasons — without much success, though his two-stroke single attracted the attention of Chris Lavery (then managing the affairs of CZ rider — later champion — Joel Robert and now, in the 1980s, doing the same job on behalf of Yamaha star Neil Hudson).

Both Joel Robert and Torsten Hallman of Sweden were impressed by the 200 lb 25 bhp Suzuki. Hallman went so far as to say that it was faster than his 1966 championship-winning Husqvarna.

Joel Robert of Belgium, winner of more world championships in moto-cross than any other rider. Three wins were on Suzukis in successive years, 1970, '71, '72, in the 250cc class.

Hugh Anderson checks his 250 Suzuki at the East German Moto-cross GP in 1967. Suzuki have won more world championships than any other make, Japanese or European, over the years from 1952.

Suzuki let it be known they were anxious to recruit top riders: but finally it was Kosimo alone who turned up on the Suzuki in 1967 to dispute a few events with the iron men of Europe. A story illustrating the gulf that exists between Oriental and Westerner in motorcycling matters is told by Jeff Clew in his book *Suzuki*. Poor Kosimo broke down in several races. On each occasion it was, according to the Japanese, 'ignition failure' that brought him low. Ignition failure it was — but caused by the frame breaking, which snapped the connection between coil and generator. Ignition failure may be admitted by the Japanese; but not, apparently, failure of the frame.

In 1968 Suzuki took on Olle Peterson of Sweden, their first official European rider, who made an encouraging beginning with second place in the Belgian GP. Code named RH, his 180 lb flyer put out about 35 bhp, was startlingly fast and reasonably controllable. Peterson thereafter fell by the wayside on the championship trail. The enormous power of the Suzuki was very clear and this,

together with the big money on offer, persuaded Sylvain Geboers and Robert to join Peterson in Suzuki's 1970 team. From that time world moto-cross, in 125 and 500 classes particularly, has been dominated by the Japanese factories, with Yamaha and finally Honda joining Suzuki. Among the 250s some leading European makers, such as KTM and Husqvarna, have fought back hard. But generally, with top-class riders like Robert and Geboers and Roger de Coster to complement the Japanese lead in technology, there was very little that could be done by the much smaller firms. Power and lightness were twin objectives for Suzuki and Yamaha. At one time the success they achieved in these areas even threatened riders' safety, and so the FIM stepped in to insist on prescribed minimum weights for each of the main cc classes. After this, the constructors' attention was focussed on improving suspension, to keep the lightweight chassis in more regular contact with terra firma.

Yamaha scored here. The company has explained that their monoshock, first known as monocross, suspension began

above
World champion Harry Everts of Belgium leads the field on his water-cooled Suzuki. Number 3 is Gaston Rahier on a Yamaha, number 1 is Akira Watanabe (Suzuki).

opposite
Speed . . . Moto-cross is second only to road-racing in spectator appeal.

131

Aberration of the 1970s. Long after two-strokes had shown they could produce more power, at less weight, than the four-strokes, a development of the XT500 Yamaha was seen in action in world moto-cross. This 1977 machine was prepared by Torsten Hallman and Sten Lundin, former world champions, for use by Bengt Aberg, another — you've guessed — world champion.

Brad Lackay of the USA on a 1977 Honda. He partnered Graham Noyce in European moto-cross and has been runner-up in the 500cc class on two occasions.

as the brain child of an independent engineer, Lucien Tilkens. Yamaha adopted the system, and eventually revolutionized the world of moto-cross suspension with it. In 1973 the factory Yamaha moto-crosser was equipped with monoshock, and Haken Andersson won the world 250 championship on it: this was the signal for Suzuki and Honda to shake up *their* ideas on suspension, which for too long had followed the conventional pivoted-fork twin spring unit setup. These days, of course, the monoshock system is a commonplace in road-racing and moto-cross, and one or

two of the Japanese firms employ it on their road machines too.

The Yamaha monoshock was first seen on Andersson's 250 in the 1973 Belgian GP at Wuust Wezel, near Antwerp. Tilkens' design had the rear subframe of the Yamaha pivoting at the normal place, just aft of the gearbox, but movement was controlled by a single spring unit running from under the seat up to the steering head. Tilkens worked on the premise that too much attention was devoted to engine power, which anyway was largely wasted because it was not translated into effective traction, the rear wheel being more often than not airborne.

Tilkens worked from home, in a workshop beneath his house. He was for many years involved with the technical committee of the motorcycle federation of Belgium, and the job brought him in touch with Geboers and de Coster, riding for the Czech firm of CZ, then the leaders in moto-cross. The newly developed CZ 400 was proving unpopular with the Belgian riders because of its higher power output, compared with the earlier 360, which made it decidedly more difficult to control. Tilkens suggested that the 400 would be radically improved by a stiffer rear frame controlled by a single spring. With this arrangement, he argued, much of the flexing and consequent general instability associated with the conventional setup, which relied for strength on the stiffness of the rear fork alone, would disappear. But his ideas were shelved. CZ were still ahead of the opposition, such as Greeves and Husqvarna and

left
Yamaha YZ400 of 1979/80, which was based on the YZ250 of '78 and gave about 45 bhp. Double world champion Heikki Mikkola's works machine contributed other improvements to the 400 . . . apparently minor, yet telling, changes such as the move to a front-fork castor angle of 60° 30', from the previous 59° 30', in order to 'slow' steering reaction. Length of the rear alloy sub-frame was increased by 40mm to 500mm, and rear-wheel travel to 265mm from 250mm. The rear shock absorber was redesigned, to incorporate more light alloy and finning, in the interests of better heat dissipation.

Maico, who were themselves handicapped to more or less the same degree as CZ with equally hide-bound ideas. Finally Tilkens felt he had to go ahead on his own. The result was a monoshock CZ which was ridden by his son, Guy, in a 1972 national event in Belgium. Later he built a chassis entirely to his own design, installed a Suzuki engine, and had the gratifying experience of seeing the bike win its first race. De Coster and Geboers had moved to ride for Suzuki by this time, and Tilkens was engaged to improve the works Suzukis. A few weeks later he received overtures from Yamaha who (presumably) made one of those offers mortals cannot refuse.

Within a year of going to work for

above
Kawasaki 100 moto-crosser of 1977 — a 90cc, 50 x 50.6mm two-stroke with six-speed gearbox.

left
Roger de Coster, five times a world champion, at winning speed on his Suzuki.

above
Four-stroke Hondas made effective moto-cross machines at sub-international level.

opposite
Graham Noyce on his 500 two-stroke Honda. He was world moto-cross champion in 1979, the first Englishman to take a world moto-cross title since Jeff Smith in 1965.

Yamaha, Tilkens had the monoshock frame as a standard fixture in the factory moto-crossers, Mick Andrews' trials bike, and the road-race machines. He has said that his system, because of the length of the damper, allows softer suspension and more rear-wheel movement. The rigidity of the rear sub-frame gives better handling under power. The improved weight transfer of the damper, into for-and-aft rather than vertical energy, means better control during braking. And of course it is easier to 'tune' the single unit, for differing conditions, than two units. Originally the monoshock dampers were manufactured by de Carbon, then later by Yamaha, by agreement with de Carbon, with gas, oil and coil springs.

Suzuki took six world championships in the four years from 1970; and two more, in the 500 class, up to 1980. Yamaha, after Andersson's 250 win in 1973, have had to share 500 class wins with Honda, at two apiece. In the 125 class Suzuki have not been dislodged from the lead in six years. Two-strokes, now usually water-cooled, have ruled, even in the case of Honda, who came to the world championship in 1978 and succeeded the following year with Graham Noyce of the UK, and in 1980 with André Malherbe of Belgium. Kawasaki too have produced motocrossers of outstanding quality; where they have trailed is in finding riders of comparable quality to the men engaged by the other firms.

Trials

After moto-cross, trials. The Japanese were late in coming to trials because they considered that there was too little advertising or sales potential in the feet-up game. Trials riding is not a spectacular sport. Nobody goes very fast. Nobody is at great risk. On television, moto-cross has an immediate appeal to millions who have no specialist knowledge of motorcycling but like to see exciting, competitive sport — especially if the exponents of the sport are covered in mud, as moto-cross men often are, and there is a near-certainty of collision and consequent damage to humans or machinery. The most resourceful TV director would be hard put to inject drama into the average trials scenario, which has the players teetering along on a motorcycle at nil mph. Trials riding calls for a good sense of balance, finesse in brake and throttle control, and deep concentration; it puts no premium on aggression and inordinate physical fitness, as in moto-cross. Moto-cross riders tend to go in for trials as a mild form of training in the off-season; and when they become too old to bear the knocks and strains of the hard game, they may round off a competition career in trials.

Trials take place during winter and autumn, and thus in bad weather, and in inaccessible parts of the countryside. Since the object of the sport is to test the ability of the rider in the worst possible terrain — in woods (leaf mould and tree roots), ravines, river beds (with and without river), hills, sand and mud — getting to any of the places where competitors are in action poses problems for would-be spectators blessed with nothing more handy than a car or an ordinary, road-going motorcycle. For this reason the only people likely to be present at some of the more remote 'observed sections' will be one or two competitors, an official, and at most a half-dozen red-nosed, feet-stamping onlookers equipped with flasks, picnic lunches and, usually, a ribald line in informed comment. It was only when trials, or something akin to trials, took the interest of paying customers in the USA that the Japanese began to see the commercial sense in manufacturing and selling bikes designed for the business.

Trials began in Britain before the 1914-18 war. If a date has to be given to the first event bearing resemblance to the modern trial, it would be 1911, when employees of the Scott motorcycle works

left
Another early trials Suzuki, a modifed K11 80 ridden by I.A. Hillier in the Scottish Six Days Trial.

left
RL250 Beamish Suzuki with lightweight frame in Reynolds 531 tubing. Suzukis imported from the USA and modified to Beamish specification sold well in the UK, rising to a peak of 1,300 in 1978.

opposite
Debbie Evans of the USA, who as a teenager finished higher in the Scottish SDT than any woman competitor had managed in 40 years. Miss Evans entertained onlookers and unnerved male competitors by doing hand-stands on the saddle of her unsupported 175 Yamaha during off-duty moments in the 1978 trial.

137

left
Two-fifty trials Suzukis make
light of sidecar haulage. Alan
Cobb on the bike, D. Harmer
in the 'chair'.

in Shipley, Yorkshire, organized one-day events for themselves on bikes in the Yorkshire dales. Two years later the first of the annual International Six Days Trials was held in Britain. For 50 years after that, with time out for world wars, trials continued as a pleasant weekend outing for sporting clubmen in Britain: on Saturdays for the national trade-supported events; and on Sundays for the innumerable minor club or Centre (groups of clubs) meetings. The trade riders might be employed by a motorcycle factory as testers or engineers. Sometimes they would undertake preparation of the trials bikes they used, which usually were no more than slightly developed versions of the models sold, in small numbers, to the public. Occasionally the riders were motorcycle dealers with a working arrangement with a manufacturer who seldom did more than supply a bike free of charge.

There was little money in trials for the riders. Bike on a trailer, they would motor the length of the country (petrol paid for), stay in a middling hotel (perhaps both before and after the event; paid for), and have a day's good sport. They were amateurs, or very near to it. Winners of the more important events, men and machines, were publicized in reports and

opposite
Akano Hyatori on a 200
Honda in a Scott Trial . . . the
ultimate test of stamina for
man and machine.

advertisements in the weekly magazines of the period, when they were of course seen only by the 'converted', most of whom would not in any case be interested in buying a motorcycle for sporting purposes. It was very pleasant, and cosy: the non-motorcycling public had no idea that trials existed (whereas in all probability they had heard of road-racing).

Then in the early 1960s trials spread to the continent of Europe. The pace quickened. Inter-nation rivalry was no longer relegated to a once-a-year occasion at the ISDT, an event shared among the nations represented by the FIM, governing body of the sport. In 1964 the European trials championship was launched. At the same time the big, powerful four-stoke singles that had dominated so long were thinned out, to be replaced by light, agile two-strokes — though at first these, like the four-strokes, were made in British factories: Greeves and Villiers instead of

BSA and AJS and all the others. But this did not last long. The English two-strokes gave way to more highly developed bikes from Spain: Bultaco and Montesa and Ossa. And then, as the Americans began to take notice of this quaint sport, the Japanese stepped in.

Suzuki were the first, though at first it was not an 'official' interest. Ordinary Suzukis were modified by clever people like Peter Gaunt, who performed well on a much-changed K11 model, and inventive Bob Collier, whose earlier work included extraordinary specials with large car engines shoe-horned into lightweight frames. He went to another extreme with a 50cc wonder bike that demonstrated in a practical — *fairly* practical — way an oft-quoted maxim to the effect that adequate gearing can make a giant of the puniest engine. Slotting in an additional gearbox, Collier ended up with a special having 12 gears the lowest of which, with a ratio of 112 to 1, gave the diminutive Suzuki

Toshi Nishiyama in Scotland
rode a 250 Honda four-stroke
which had been shipped
direct from Japan.

'Private' 1975 Kawasaki
ridden by R.P. Salt.

Early 250 Suzuki fitted with trials tyres. In reality this pleasant two-stroke, and others of similar style, are best reserved for green lane-ing.

astounding climbing ability at the cost of over-revving at approximately 5 mph on the flat.

When Suzuki Japan turned their attention to trials, it was to satisfy Californians who had conceived a liking for desert racing. Shipments of high-performance 250s were sent to Santa Fé. Most of the bikes remained there. There

was, undeniably, a demand for cross-country bikes but it had been grossly exaggerated. The result was a glut of Suzuki RL250s in California. Finally a batch was sent to Britain at knock-down prices. Fifty were sold, at a little more than a knock-down rate but still quite reasonably, by a main Suzuki dealer, Graham Beamish. The mudguards were

fittings, they were distributed exclusively by Graham Beamish, finally under the name of Beamish-Suzuki.

But as the Japanese awoke to the interest in the sport they turned out machines designed particularly for the British market, some with 325cc engines and all with featherweight frames featuring aluminium and other weight-saving materials. These too found a ready sale, but only after further modification by Beamish, who was forced to *add* weight, in the interests of stability in the rough. Then as the stockpile dwindled in the USA, so did Japanese interest. The planners realized that sales of these highly specialized machines could never amount to more than a few hours' production time at any of the highly automated plants in Japan.

Kawasaki and the other manufacturers have kept pace with Suzuki, basing their operations in Britian, as the

altered and the engines made a little less 'peaky' with variations in combustion-chamber shape and compression ratio, and a smaller choke for the carburettor. More were ordered from America. In the years after 1974 more than 1,000 came to Britain in this way. Greatly modified, with new lightweight frames in Reynolds 531 tubing and other British-made

centre of the sport in Europe. Kawasaki took on Don Smith, a European champion, to head their early efforts. Honda went to the top man in the business, Sammy Miller, who laid down the blueprint for their first 125cc four-stroke trials models. Currently it is Yamaha who are putting most effort into the trials sphere. The Japanese interest continues . . .

Kawasaki's KL250, an ohc four-stroke trail machine with impeccable handling.

Drag racing

If one thinks of drag racing as two men on motorbikes trying to out-accelerate each other, then it must rank as the oldest form of motorcycle competition. And drag racing *is* just that: with the added refinement of split-second timing for each man to produce three-decimal-place figures of time and speed to pad out the record books and, sometimes, confuse the issue for the uninitiated. (How confuse? Because each man is separately timed, which means that if one is slow in getting away he may lose the race while putting up the better time and clocking a higher terminal speed.) A purer, not so popular, form of the art is sprinting, in which one man races against the clock. This was the way favoured by the British

for many years until drag racing was imported from the USA in the early 1960s.

Then what was previously a very private pastime, in which competitors awaiting their turn, family and friends formed the meagre audience, was transformed into a major sporting spectacle attracting thousands of paying visitors. With the infusion of money, sprinting-turned-drag racing in Britain became much more competitive. Performances improved dramatically. The bikes were bigger, noisier and more powerful than the sprinters. A drag meeting had something of a carnival air.

America was where it began, in the late 1940s, when teenagers and ex-GIs with plenty of money raced hot-rod Harleys and Indians on the highway. It was usually fairly dangerous, almost always illegal. ('Drag' as a title possibly

derives from stories of dragging the tail of a coat as a challenge; another suggestion is that it has something to do with drag hunting, in which the scent of a quarry is dragged on the earth for hounds to follow.) Drag racing, at first, was merely a part of the postwar scene in middle-western America, having much to do with the infamous 'chicken runs' portrayed in films such as East of Eden — where nerve was tested to a suicidal limit. Finally police whose job it was to find the drags, and not infrequently attend to some of the casualties when they were too late, decided that the contests would have to be regularized.

Within a couple of years drag racing was transformed, perhaps so quickly in exaggerated reaction against the former anything-goes atmosphere. Safety became paramount. The rule book swelled to cover any eventuality. If a competitor wanted to fit a lefthand-thread bolt in some particular corner, there were a few lines in the rule book to pronounce on the 'legality' or otherwise of such an idea . . . and instructions on the precise way to set about fitting the bolt, should it be acceptable. This is how the sport developed in America. In Britain and Holland, the main centres for drag racing in Europe, there is still some way to go before meetings are run with US-style professionalism.

The very fastest drag bikes are the biggest — which may be unsurprising but is true, not because in principle big engines mean better times, but as a result of one or two of the leading men investing much skill and money into building mammoth-engined specials. Indeed, many 500 and 750cc dragsters are faster

Double-engined monster at the start of a drag session.

Drag ace Henk Vink doing a 17.48-second s/s kilometre, with a terminal speed of 158mph.

than multi-engine devices, in which excessive weight and doubtful handling can cancel the power advantage. What the drag exponent craves above all else is plenty of torque. A narrow power band, however impressive, high revs and frequent gear-changing are not for him. In fact he will be happy to do without any gear changing at all, if it can be arranged.

Even the newest recruit to the sport supports the view that gear-changing is time spent going nowhere. In all-out acceleration the fewer gear changes there are the better. Hence *no* gear-changing as an ideal: which of course places a high premium on sheer power to haul the dragster from zero mph to perhaps 180 on a single ratio. There is a saying: *Drop the clutch and ride the wheel . . .*

Dragsters are split into several categories. There are production machines, as used on the road, with these being refined into another class, Pro Stock, for highly developed petrol-burning bikes

that outwardly resemble production machinery but put out much more power (150 bhp from 1,000cc being a not uncommon figure, for example). The other main division covers competition bikes where 'anything goes' for machines built purely for drag purposes, with engines up to 3,500cc in normally aspirated form and 2,000cc when supercharged. Machines running on special fuels are known, rather tamely, as 'fuel' bikes while the petrol-burners have the more colourful label, 'gassers'. An unconscious case of compensation, perhaps.

In the USA, in a pre-Japanese age, the favourite motorcycle for drags was the Harley-Davidson; and even years later, in 1977, these old-fashioned overhead-valve big twins, sometimes doubled up, were as fast as the best of the Japanese-powered devices. The Triumph vertical twin was another favourite unit in the USA, and of course in Britain. By the 1970s it was almost routine for engines to

146

nitromethane fuel mix) for urge. In Europe, Holland and Britain mainly, the chosen way has been supercharging and use of 'nitro' — four-stroke engines are reckoned to be more suitable cases for treatment than two-stroke multis; thus most record-breaking dragsters are Honda and Kawasaki based.

The first man to get into the 'sevens' for the quarter-mile was Ross Collins of Los Angeles riding a lengthy Honda projectile. Mr Collins spent almost 20,000 dollars in building his 10 ft-long dragster, powered by three Honda 4s, each one enlarged to 1,100cc. With this 850 lb baby — christened *Atchison, Topeka and Santa Fé* — Collins got down to 7.86 seconds at Ontario, California. Shortly afterwards, at a meeting in Ohio, he and bike parted company at about 160 mph, which signalled the end of *Atchison, Topeka and Santa Fé* and, almost, of Collins. Later he built another, slightly shorter device (only two Hondas this time) on which he clocked a 7.62 seconds, 199.55 mph run in 1978.

In Europe the main venue for drag racing is the UK's Santa Pod raceway, opened in 1966. Drag stars from Holland, like Henk Vink, and the USA come to race against the best of the British, like John Hobbs and Pat Higham (Suzuki), among others. Vink is the main Kawasaki importer for Holland, and a man of great drive and determination. His early business success enabled him to finance his other obsession, drag racing, which began in 1967 when he rode a production Kawasaki 500 triple at the first of the Anglo-Dutch drag contests. Later he installed a 750-3, quickened with factory components, in a lightweight chassis which brought him times in the low 'tens'. In 1972/73 Vink asked his friend Phil Manzano, a tuning expert and rider running a small workshop in Kent, to build a supercharged 903cc Z1 Kawasaki. Despite the power shortages imposed by the then three-day working week, Manzano and one of Vink's mechanics turned out in three months a Wade-supercharged, nitro-burning, 300 bhp Kawasaki that was to take Vink to the world's standing-start kilometre record with a one-way 16.01 seconds time and an average speed over the last quarter-mile of 197 mph. The strength of the Kawasaki engine is shown by Manzano/Vink statistics which record that the same, standard crank assembly was used in a series of engines built over five years in the 1970s, during which time 60 sets of specially forged pistons were 'consumed'.

be doubled. For a Triumph, usually bored out to 850cc to start with, this meant ending up with a not-unwieldy four of around 1,700cc. As Honda, and later Kawasaki, came along, the same procedure was followed, when the final setup would comprise a power plant of eight cylinders, totalling about 2,000cc, keeping two heavily burdened wheels at least 7 feet apart. Not that constructors anxious to take advantage of Japanese machinery had to wait for the big four-stroke fours to arrive: in the USA especially, Kawasaki's two-stroke triple, usually the 750cc H2 version, proved very popular. In Britain John Lloyd had three of these engines in tandem — nine cylinders, 2,250cc — and as late as 1980 was setting sub-9 second runs (terminal speed 153 mph) over the quarter-mile in the gasser class. But in the main it has been Americans who have favoured really big jobs relying on vast torque (augmented by draughts of 90-95 per cent

Tomorrow's Classics

Does any Japanese motorcycle qualify as 'classic'? What, anyway, is a classic motorcycle? The dictionary says, of classic: 'of the first class; of allowed excellence', which is useful, but then rambles on about classic being synonymous with simple, harmonious and proportioned — criteria applicable to very few motorcycles, and those few are on the whole too uninteresting to merit further attention. The Vintage Motor Cycle Club (it seems reasonable to equate, to some extent at least, vintage with classic) says, in its wordy way: '. . . engineering skill and vision of those who built the pioneer machines . . . a new generation can marvel . . . an older generation view with nostalgia . . .'. A down-to-earth, ultimately more valid definition might have it that a classic motorcycle was one which gained in monetary value, instead of depreciating, as it aged. The Vintage Club has 25 years back from contemporary view as the cut-off for admission to vintage arcadia, which effectively rules out most of the Japanese ware. But the age test is a very arbitrary one, and not invariably reliable. There is much pre-World War Two machinery that is undeserving of any status more exalted than 'boring' or 'cheap' — the sort of motorcycle an experienced chronicler has dubbed, with awful finality, 'grey porridge'. To name a few names — and why not? — such once familiar British workhorses as the late 1930s Red Panther 250 (and most Panthers of any age, any size), assorted BSA 250s, and big side-valve Ariels, qualify as very lumpy porridge. Yet — keeping to English marques — many post-1955 motorcycles, among them the Featherbed-frame Nortons, the last of the Vincent vee-twins, and the Gold Star BSA, now change hands at figures well above their original list price. These are, incontrovertibly, classic motorcycles.

Rereading the last few lines, I realize that another, possibly essential condition in a motorcycle attaining classis status is to have the maker go out of business, preferably in painful circumstances and leaving a great many unpaid bills. At present none of the Japanese Big Four has given any indication of meeting this requirement though owners of Honda motorcycles with an eye to future auction prices may take heart from the association of Japan's leading motorcycle company with British Leyland, which, a cynic might suggest, can only prove beneficial in this respect.

Ever-increasing sales and copper-bottomed solvency notwithstanding, the Japanese have produced, and continue to produce, occasional classic motorcycles. Following, in no particular order of age or specification, are details of some models that fulfil my definition of the genre.

Honda CB92

In 1962 I rode a 125 Honda CB92; it was a time when Hondas were very thin on the (European) ground. In Britain only four of the 23 models then manufactured by Honda were available. This 125 was the most interesting of the four. It was, if such can be said of any Honda, hand-made: not literally, of course, or even in the sense that the term might be used in connection with a small to middling manufacturer; but on the Honda scale, by comparison with the millions of bikes that were to follow, for the CB92 such a description is not wholly inaccurate. In 1962 it was very expensive on a cash/cc basis, which, incidentally, even at this early date should have given the lie to those diehards who claimed that the Japanese, subsisting on their legendary bowl of rice a day, could afford to sell cheap. But being expensive is another vital constituent of the would-be classic. No matter if he becomes violently unhappy with his new machine minutes after leaving the showroom, a stubborn owner, having parted with too much money, will steadfastly refuse to blackguard his toy. Instead he will praise it to the skies, do anything rather than admit that he has been a fool when he could — should — have taken sensible advice and invested in a perfectly adequate, if more humble device at a fraction of his present expense. Thus a myth is born. Only in exceptional circumstances is the truth allowed to slip out. It is this understandable self deception that has saved the bacon, for a time anyway, of engineering 'classics' as unworthy and various as many Aston Martin motor cars and the Concorde aeroplane. In the case of the 125 Benly, though, the selling price was entirely reasonable. The quality throughout was very high. The late Donald Crawford, talking about his Benly, had this to say:

'To me the CB92 has a distinctly purposeful and functional appearance. It has an aura of speed about it which is accentuated by the angular fuel tank and the huge brakes. The front of the bike takes your eye at once, the 8in-diameter twin-leading-shoe brake indicating high stopping potential in keeping with the rated performance of the engine. The tyre size is 2.75 x 18in. The forks are box-patterned and immensely strong; they are not prone to whip as are some of the telescopic type. The suspension is leading-link and absorbs most road shocks without notice. Rear suspension is very different, and deserves criticism. The damper units are adjustable to three positions but the springs are so strong that even on the softest setting they remain too 'hard'. This, coupled with the tyres, gives a most uncomfortable feeling on occasions, particularly in the wet.

The speedometer reads to 90 mph and shows mileage down to tenths. Just above the speedo is a red neutral-indicator light. An optional extra is a rev-counter calibrated to 14,000 rpm. The rev-counter drive is a standard fitting situated on the offside of the cylinder head and working off the camshaft. Matched against the rev-counter, the speedometer is commendably accurate. Road speed in top gear is 8 mph per 1,000 rpm.

Although a kickstarter is fitted, I use the electric starter every time, and not once has it failed, in spite of being subject to much use in cold weather. The electric system is six-volt with a dynamo of special design which regulates the maximum output automatically, so that the current output does not correspond directly with engine revs, and therefore no special voltage regulator is required.

Donald Crawford on his CB92 Sports 125 Honda: a jewel of a motorbike.

149

The cylinders are inclined at 60° and have bore and stroke of 44 x 41mm. Cylinder walls are finished with a honing machine to an accuracy of within 0.01mm of taper or out of round. High-silicon aluminium alloy is used for the pistons, which have domed crowns and two compression rings and one oil ring. The output of the engine is 15 bhp at 10,500 rpm, and maximum speed is 81 mph. The high maximum rpm figure indicates that the power band is high too; in fact real power does not arrive until at least 6,000 rpm, when things start to happen very suddenly. Third is a particularly flexible gear and I have covered many miles in this ratio at a steady 70 mph, when the rpm reading is 11,000 and the engine feels as if it could maintain those revs indefinitely. After a couple of miles at 11,000 in third I was trying to ignore an insistent vision of four tiny valves being forced up and down at a dizzying rate; but there was never a hint of trouble and I now have utmost confidence in the motor.

Gearbox capacity is only 1½ pints. It is advisable to check the oil fairly frequently. Oil is pressure-fed around the engine by means of a pump on the right side of the crankcase. This plunger-type device, cam operated by the crankshaft, drives lubricant through a submerged oil screen and into a chamber located on the crankcase and then to the oil gallery and vital parts. A centrifugal filter on the end of the crankshaft winnows impurities from the oil; a magnetic drain plug traps any metal particles getting past the filter. So far no production racing 125 Honda is available, but the standard CB92 can be fitted with a race conversion kit. It includes special pistons, racing camshaft and beautifully chromed megaphones as well as an assortment of other items such as jets, gaskets and alternate sprockets. This conversion makes the Sports Benly an excellent introduction to racing. In this year's TT a converted CB92 finished well up the field, indicating that it is no sluggard even when matched against factory machines.'

Touring version of the Crawford 125 was larger, flabbier and no doubt rather more sensible than the CB Benly.

150

Honda CB72 and CB73

Other Hondas worthy of nomination as classics include the 250cc CB72 and its very close relation, the CB73 of 305cc. (The Japanese indicated early that they were free of traditional notions about engine capacity, which in the West had followed a 125, 250, 350, and 500cc pattern for many years; soon we were to become accustomed to, first, 305, then 185, and 450, and 425). The impression these two Hondas made in a European market used, by way of 250s, to nothing more exciting than BSA Group and AMC singles and the motley, but mainly mediocre, selection of small-make models powered by the ubiquitous 18 bhp Villiers two-stroke twin, was electric.

I recall writing, in 1965, that the 305 was beautifully made, lighter and probably a shade less rugged than a BMW, but displaying attention to detail and workmanship and styling of the main components, such as the engine and the brakes, that had much in common with the Teutonic marvel. (In those days we tended to be more impressionable than now, and the BMW, while something of a lump, undeniably was nicely made).

The 305's overhead-camshaft parallel twin engine had a light-alloy cylinder barrel and head and was canted forward, forming an integral part of the machine; in other words, there was no front down tube — the engine did not sit in a frame, in the conventional way. Bore and stroke Enthusiasts for Japanese 'vintage' machinery being as intemperate and prejudiced as any other nostalgia-ridden fanatic, there tends to be talk of the CB72 of the early 1960s being the last 'real' Honda. More prosaic facts are that 20 years ago Japanese bikes in Europe and the USA were comparatively rare and had greater impact than later. But the ohc Honda was a very fine motorcycle in 1961, a match for BSAs, BMWs, flat-head Harleys and all the others of much greater capacity.

were 'oversquare' at 60 x 54mm, and the crankshaft was carried in two ball and two roller bearings, with roller bearings for the big ends. The compression ratio was 9.5:1, petrol was fed via two Keihin carburettors and maximum power was around 27 bhp. Oil, 3¾ pints, was carried in the sump, and the clutch ran in oil, with primary drive by chain. Gear ratios were 17.48, 10.42, 7.33 and 6.27 to 1. An electric starter was mounted on the crankcase and energized by a 9ah battery, with power from a 12 volt alternator. Tyres and wheels were typically Honda in being slender, 2.75in and 3.00in, with 8in-diameter twin-leading-shoe brakes carried in full-width hubs.

The unfamiliar electric starter was not an unmixed blessing. When the 305's engine was cold, considerable finesse was called for in setting the air slides and the throttle twistgrip; and even after several hundred miles with a Honda a new owner might still be uncertain of the best way of getting over cold-starting blues. The slides had to be opened just a little, say to a third of their travel, and the twistgrip set a hair's breadth off the stop before the button — or (for the first start of the day) the kickstart — was used. The engine would start, after several near misses, and then had to be left to warm up at moderate revs — about 2,000 — with no attempt made to open the throttle until a working temperature was attained.

Ignition and lighting arrangements were complicated. A control key was housed in the top rear corner of the nearside air-cleaner/toolkit cover . . . for many years Honda remained out of step among the Japanese manufacturers in insisting on tucking ignition controls away in some awkward-to-reach spot when it would have been more convenient, for the owner, to have them on the fork crown. The key had three positions: off; ignition and lights on; parking lights only. With the key in the second position a knob on the headlamp had to be turned for the side and rear lights, and moving the dip switch on the left of the headlamp did not affect those lights. But on turning the knob in the opposite direction, the side light could be augmented by the main beam, lowered or on high, on operation of the dip switch to the extremes of travel; then the middle position for the dip would bring in the side light alone. In the third position, the side and rear lights being on irrespective of the headlamp knob, the key could be withdrawn. For new motorcyclists brought up in the dawn of the electronics age this rigmarole was unremarkable; but to any long-time

rider used to more spartan arangements it was a mind-blowing experience.

A natural sequence in getting away on a 305 was to take the engine to 6,000 rpm in the gears before getting into top, which worked out at 20 mph for first gear, 40 for second and 55 in third. In top gear 6,000 rpm equalled 70, and that was an easy, pleasant cruising speed . . . until realization would hit the rider that approximately 3,500 revs were going to waste, and the rpm rate would be upped to 'seven' — about 80 — for motorway cruising. At this speed a 305 vibrated very little — not enough to worry the rider, and certainly not a rider accustomed to British vertical twin vibes, though a pillion passenger could find the high-frequency *zizz* a little obtrusive. Third gear was good for 85 mph, when the rev-counter reading was a full 9,000. All gear changes at high engine speeds were fast and clean. The Honda's gearbox did its work efficiently if with no particular delicacy: the action, when one had time to be aware of it, was firm, a little sticky, although the movement required was small. A slightly crouched rider could attain a genuine speed of 95 mph, when engine revs were reaching 8,800 . . . there could be uncertainty on this score, for 305s were susceptible to a little swing on the rev-counter needle at high revs which made if difficult to establish a firm reading. Alternatively, a 305 *could* be kept down to 30 mph in top gear — 2,250 rpm — though doing so smacked of cruelty to dumb engines. Steering was adequate, with no shake or shimmy on fast bends and no undue lightness at the steering head. If this appears to be a rather flat way of describing a 305's handling, it is nevertheless accurate; the Honda did not, for instance, measure up to the 'standard' of the time, the Norton Featherbed, but was, as indicated, entirely reasonable. Adequate.

I concluded a road-test report on the 305 in this way:

'Having had our motorcycling notions formalized during a period when British motorcycles set the fashion, and as a consequence nursing even to this day a conviction that 100 mph should be attained at the cost of no more than 6,000 revs per minute ['seven-five' and similar expressions were exclusive to racing men and racers could, and frequently had to, dismantle their engines after every meeting] we approached Hondas without in any sense being 'sold' on them. But they, and the 305 in particular, have persuaded us that there is nothing fundamentally indecent in trundling along at an easy 8,000 rpm. It may be a little un-British, but that is another matter.'

Honda CB450

The final Honda to be included in this section is the 450cc double-overhead-camshaft twin that was exported to Europe in 1966 and widely advertised, for reasons never made clear, as the black bomber. There was, presumably, some lunatic intention on the part of the advertising copy writer to stir in the motorcycling public dormant memories of one-time world heavyweight boxing champion Joe Louis. But he had been the Brown Bomber; had not, in 1966, been news in over 10 years; and in any case had been as large and heavy as any of the unfortunates who disputed title matters with him — all of which made the 'black bomber' advertisements showing diminutive (!) 450 v Vincent 1000, with statistics overwhelmingly in favour of the Honda, doubly mystifying. However, they helped the bike to make a decided impression and stirred purists (usually boxing and/or Vincent fans) to write furious letters to the motorcycle press. It was, as things turned out, a textbook case of oversell. Coming so soon after the enormously successful Honda two-fifty twins, which had been shown to be rather more than a match for most European motorcycles of double their capacity, the 450 made its debut in a welter of publicity of the black bomber kind that fuelled expectations of staggering performance. The 450 performed very creditably but could only disappoint

those enthusiasts who had been readying themselves for a maximum speed of 120-plus and breathtaking acceleration.

Selling at £560 in those far-off days, the 43 bhp twin had two overhead camshafts (contrary to established sohc practice on the smaller road-going bikes) driven by an extraordinary long chain taken from a sprocket on the crankshaft and careering up and over the camshafts, guided and restrained by more than half a dozen idlers and several tensioners. The valves were actuated by rocker arms below the cams, the arms pivoting on eccentrically mounted spindles at one end (eccentric to allow for valve-clearance adjustment), receiving the motion of the cam at about the middle of their length and striking the valve stem with the other end. For each valve another arm, similar to the rocker but shorter and mounted below it, curved up and had its free, forked end against the under surface of a collar at the valve tip; this arm pivoted from a torsion bar mounted across the cylinder head. There were four bars, one for each valve, and it was almost certainly a unique method of returning motorcycle valves to their seats. The cylinder head had a cast-iron skull in its alloy jacket. The carburettors, Keihins as on all Hondas of the time, were of 32mm choke and unusual in a motorcycling context in being of constant-vacuum type in which there is a slide in the intake, its movement controlled by a vacuum in the manifold. When the manual throttle, of butterfly type, was opened, the vacuum

Black bomber of 1965. Superb engine, surprising — but not startling — performance: the CB450 was a thoroughbred but had something of the workhorse about it, for examples have clocked 100,000 miles with only routine replacements.

153

M.R. Wigan, enthusiastic amateur racer and long-distance tourist, who owned a CB450 before moving on to one of the first CB750s to arrive in the UK. He christened it Phidippedes and remained violently pro-Honda until 1972 when Kawasaki's Z1-900 — a little faster, more race-worthy — came along.

right
It has been axiomatic that bikes for hauling sidecars should be big, beefy and preferably low-revving 'sloggers'. Imagine the horror (not least at Honda HQ) when it was suggested that a 10,000 rpm CB450 should be pressed into sidecar service. But the E.O. Blacknell sidecar firm of Nottingham devised fittings and clamps for the strictly solo-only frame of the 450 and the union, as experienced C.E. Allen, Vintage Club founder, testified, was extremely satisfactory. It would even 'slog', he said.

increased and raised the slide further, to keep the vacuum constant.

Another change from normal Honda chain drive was primary power transmission by straight-cut gears, all gears being indirect. By British standards, of course, overall gearing was ludicrously low for a (near) five-hundred at 18.58. 10.70, 7.97 and 6.96 to 1. But the CB450 thrived on high revs, like other Hondas, and 70 mph in top gear worked out to approximately 6,500 rpm. The frame, another change from usual practice, was of full cradle type with large-diameter, single down tube, the rear pivoted fork being carried between wide-spaced tubes in approved British fashion. There was coil ignition, with an ac generator charging a 12 volt battery mounted under the dual-seat, and electric starting. I rode one of the first to arrive in the UK and said, loftily, that at a time when few new models were being introduced the new 450 brought back some of the kicks to motorcycling. Acceleration was average for a 500 — at least, to about 70 mph, from which point it would have taken a very good 500 to hold the 450. I estimated that 104-106 mph would be possible in calm conditions (my test trip was undertaken on a day of gusting wind), possibly more if one were flat on the tank.

That was the last I saw of a Honda 450, so far as any chance of riding one was concerned, for a long time. However M. R. Wigan of the Road Research Laboratory, who was uncommonly quick off the mark in acquiring new motorcycles (and is more fully described in another part of this section), bought a black bomber within weeks of its introduction and was unkind enough to make me envious with his dispassionate recital of the delights,

and some few disappointments, of 450 ownership. Said Wigan:

'The immediate impression is of a stumpy, square shape dominated by an enormous, gleaming alloy engine. [This was 1967, remember, when a 450 with some excess finning would pass for "enormous".] The impression of weight is well founded, because total weight is close to 500 lbs. The reason for this in a 444cc bike is soon discovered: the frame is of heavy-gauge tubing, the seat is very heavy, and the large tank is in thick gauge

154

steel, as are the mudguards. It would not be difficult for a determined customer to shave off a very considerable proportion of this mass. The components that add so much to the weight do a good job for the touring rider. The mudguarding is efficient, the seat is extremely comfortable for two people for extended journeys, and the large tank is very well designed, holding 3½ gallons of which never a drop slops out from the cap, under the most severe provocation [a modest reference, no doubt, to the Doctor's vivid riding style]. The frame is over-solid — it would be improved by a slightly extended pivoted fork and a modified treatment of the attachment. Both these changes are rumoured to have been incorporated in the 450 Mk 2, on sale in America.

The front suspension is excellent, though it is improved by the use of Castrol Shockol in the forks — Castrol appears better suited to the English climate than are the standard Honda brands. The rear suspension is poor; earlier units were moderately likely to seize up after 5-6,000 miles. On my 450, standard Girling units, fitted with 110 lb springs an inch or so too long (to give a pre-load of about 70 lb), transformed the handling and gave me great confidence when running in a production race in a streaming cloudburst. The only remaining criticism of the handling characteristics is a simple consequence of the great weight carried high, which makes rapid changes of direction through an S bend hard work. The steering is very good. I fitted a hydraulic steering damper before venturing on to the track, but found that it was unnecessary. The brakes are typical of Honda. The front is very powerful, entirely adequate for fast travel with a heavy load aboard. However there is a definite reduction in efficiency in racing conditions, a not un-expected result from using road-going brake lining materials. One would be well advised not to fit 'green' linings for road work, as there would then be a notable excess of braking power, and in wet or otherwise inclement con-ditions this could be embarrassing for most riders. [These comments have "dated", of course, in view of the wide-spread adoption of powerful disc braking in the early 1970s.]

Engine braking is very strong, and the com-bination of considerable torque at low rpm, a sensitive clutch and this engine braking could be expected to give rise to a very jerky and uncomfortable ride in town. That is does not do so is due to the very clean carburation with the new constant-vacuum instruments.

The features on the 450 causing most distrust among the unconverted are the long timing chain and the torsion-bar valve springs. No trouble of any kind was encountered with these components, and the compression measured at 15,000 miles was the same as at 3,000. No sign of valve float was ever noticed. I am convinced that the torsion-bar system is as close to a perfect set up as we are likely to achieve at a reasonable cost. Even desmo-dromic systems require more attention, either in assembly (cost) or in maintenance (expens-ive consequences of inadequate care in wear compensation).

The engine is, unfortunately, very tall and looks rather like a Manx motor, with the massive cooling fins on the head. The double overhead camshaft set up has been very well developed to take full advantage of the deep breathing that is made possible by the valve-gear layout; the engine refuses to pink on 90 octane and gives a tremendous spread of torque with good economy in fuel. One criticism of the design is the pressure oil feed direct to the cam faces — it takes about two minutes to build up, and until then the lubri-cation is hardly adequate for the 700 rpm tickover. The riding habits of some owners have thus led to very rapid wear on the faces of the camshafts.

Carburation on the 450 caused Honda much embarrassment in the early days, with bikes going out with settings adjusted for the "wrong" countries. This was quickly rectified,

though, and a free quiver of jets produced in short order. When these kits are fitted, the carburation is very good throughout the speed range. My opinion is that the main jet should be 89 instead of 125, but beyond this I can find no fault with these 32mm carbs, which refuse to "fluff" no matter how brutally one snaps open the throttle. I remain impressed after many thousands of miles, and add constant-vacuum carburettors to the specification of my "ideal" bike.

When I rode the 450 in a race meeting I found it the equal of most 650 twins of British manufacture. I rode over 100 miles to the race venue, the bike in full road trim, removed panniers, number plates, raced in pouring rain and then rode home, with the panniers, and so on, refitted.

Naturally road tyres were in place, and the engine was precisely to standard order. Handling and braking were entirely safe, and I had the pleasure of finishing some way forward of the back of the field in spite of exaggerated caution owing to my lack of any alternative transport home.

The machine always started easily. At the start of the race I tried to be clever, and put the 450 in gear so that I could blip the starter button and go straight off. The flag went down and I started the motor as planned but let the clutch home too fast, before the second cylinder had caught, which made a mess of things . . . so I disengaged first gear, started the bike with one swing of the kickstarter, and still managed to sweep into the first bend ahead of the pack. As I was changing up I heard the commentator say ". . . and the Honda is off first, with its electric starter." Although people deride the use of an electric starter I have come to appreciate that it is a first-class safety feature . . . If the bike stalled, I didn't have to stop, slow down or even move my feet: one touch of the button and the engine fired immediately. The low-speed torque and good handling make it a very comfortable and effective town bike. The engine, though, ran very warm and this often worried me.

Life of the usual fast-wearing items on the 450 was rather variable. The rear chain required only one replacement, which worked out to a life of 8,000 miles. The sprocket was still serviceable. The rear tyre lasted about 4,000 miles, the front 6,000. The life of the plugs was undetermined, for I used appropriate plugs for differing conditions: always, however, at least one grade harder than recommended by the handbook, and always NGK . . . I would estimate a life of about 5,000 miles for D9H plugs.

The general impression one had in riding the 450 was of effortless performance and unfussy temperament . . . it was certainly one of the best touring and general-purpose motorcycles made.'

Kawasaki H1 500 and H2 750

Kawasaki? The H1 500cc two-stroke triple, sometimes known as Mach 3, must be included. In the late 1960s Kawasaki retired, reputation bruised, from an unsuccessful attack on the US market. They had taken a big overhead-valve twin cloned from a 650 A10 BSA — in those days the Japanese had not out-grown their legendary talents as copyists — to the States, confident that the bike would repeat the gratifying sales it had notched up on the home market. But selling a bike in a market adroitly secured against competition from abroad was one thing; pushing it in free-market conditions where a strong bias existed in favour of 'traditional' big bike manufacturers from Europe was quite another. The situation was made piquant by the Kawasaki Commander, as the A10-lookalike was known, appearing in showrooms alongside the English original; once they were out of the showroom the A10, especially in its later Rocket form, consistently outperformed the Kawasaki. So the Japanese sold Commanders at giveaway prices and sailed away to Kawasaki Heavy Industries in Kobe. There, after more than a little un-Japanese wrangling between four-stroke proponents and the 'stroker' brigade, the triple made its debut. Only of course it was not quite as clear cut as that . . . When the decision was made to go two-stroke, splits in the winning side became apparent. Two-stroke, yes — but was it to be a twin or a triple? Examples of both were constructed for evaluation. The three won. Smaller cylinders and pistons in the three were more easily cooled, with the middle cylinder getting almost as much finning as the outers. The ports, being piston-controlled, could be larger in total area than those in a twin. Finally, against all expectation, the triple was lighter.

Lubrication was based on the system devised for the firm's road-racer, and was extremely thorough. A metering pump took oil from a five-pint tank to the main bearings, thence to the big ends, with a supply going to the inlet ports for jetting on to the piston skirts: all to minimize the traditional bugbear of two-stroke seizure. An overriding control was linked to the throttle, in the manner followed by other two-stroke manufacturers, to ensure that lubrication was roughly in tune with

engine requirements, with the object not merely of avoiding oil starvation at the moment when engine pressures were at their greatest but of ensuring, equally, that the bores should not be awash with unwanted oil at tickover — conditions leading, respectively, to possible death following piston seizure, and almost certain apoplexy on the need to clear the plugs of oil-fouling. For all those fabled 'few moving parts' of Mr Day's brain-child, the two-stroke designer has had a

hard time! On early Mark 3s any tendency to plug-fouling was further reduced — though Kawasaki's PR hyperbole did not say 'reduced', of course, but insisted that plug-fouling was eliminated — by an

Kawasaki's Mach 3 on show. Deceptively ordinary-looking and smallish overall, though a trifle broad in the block, this was the motorcycle that cheerfully ruined the carefully constructed image — courtesy of Honda's CB750 — of the motorcycle as fast but smooth, safe and civilized. It achieved 750-style performance on two-thirds the capacity and had memorable road manners.

electronic ignition system that undeniably was ahead of its time. It kept the plugs happy but, inconveniently, suffered from peculiar troubles of its own in its induction coil wizardry. Within a year production Mark 3s were being sold with conventional ignition.

Large induction and transfer ports with moderate timing were features of the 60 x 58.8mm unit with its 120° cranks. Inevitably, as a three, the engine was wide. This condition was aggravated by the overhanging generator at the left-hand end and the ignition distributor at the other end. The crankshaft was dowelled into the flywheel and the main bearings assembly, an arrangement which made good engineering sense but cost the owner dear when, and if, need arose for a bearing replacement, because than he had to pay for a new crankshaft as well. The motor gave . . . very willingly . . . at least 60 bhp. Strangely, even old hands in the USA and in Europe long used to (and sceptical about) high-flying claims in the transport world appeared to accept that figure. Certainly they accepted it without question were they not such old hands as to be precluded from actually riding the thing.

Where the makers, or bedazzled observers, tended to mislead the great motorcycling public was in claims for the Kawasaki's handling, which they said was excellent because it derived from a frame that was all welded up and was, generally, a copy of the Featherbed Norton. In fact it was not up to the job of keeping 60 horses decently curbed. The bike weighed under 400 lb ready to go, had a top speed of 120 mph and would cover the standing quarter-mile in approximately 13 seconds. This, at the end of the 1960s, made the Mach 3 the fastest production motorcycle.

It was distinctly quicker than contemporary biggies like the BSA/Triumph 750s and the Norton Commando, and dented the reputation left behind by defunct Vincent vee-twins. On English roads it would out-accelerate the E Type Jaguar, motoring's kingpin of the time, and in the USA, Kawasaki's main overseas market, it was too much for the likes of Corvette and Mustang. Ironically, it is the deficient handling that secures the Mach 3 in memory: its handling allied — more accurately, in conflict — with the enormous power. The wheelbase was stretched by 2 inches from 51 inches when the bike reached production but that was not long enough to keep the front wheel in contact with the ground during quickish getaways. One adjusted to this idiosyncracy, of course. The front wheel could be left to float, or not, depending on situation and what Ernest Hemingway used to call *cojones*.

That was predictable handling; but on many occasions the Mach 3 was far from being predictable. Through corners, especially bumpy corners, where power might come in with a rush following some hiccough in the carburation department — such things happen — or an inadvertent jolt to the twistgrip, rider control could become non-existent. The Kawasaki would then take over, landing you beyond the corner possibly on the extreme margin of 'your' side of the road, or on the wrong side of the road, or conceivably not on the road at all. It made for exciting motorcycling. One of the normally ebullient American pressmen said, after trying a triple in early 1969, 'I would say that some of our more impetuous lads are going to get into difficulties with this motorcycle.' The subdued tone of those words indicated that he had found a few miles by H1 a nerve-wracking experience.

But no bike worth listing among the classics is likely to be bland, safe, economical or even ultra reliable. Contrary to the nth degree in all these respects — no, that is not quite fair; reliability, at least, was reasonable — the H1 should be assured of its place.

A year or so after the introduction of the 500 triple, heartened by its success in the USA and — to a much smaller extent — in Britain, Kawasaki followed the honoured Hollywood maxim of giving the public another helping of the same, only bigger. Roll of drums! Orchestrated squawks from all 'anti' bodies dedicated to freeing the environment from two-stroke noise and smoke, and ending two-stroke decimation of young lives! Hail, son of H1! (Name of H2, naturally.) It was a 500 triple enlarged to 750 and producing, between seize-ups and fuel stops, about 75 bhp to propel speed-addicted owners through the quarter-mile in 12 seconds dead. This one is a classic too, for roughly the reasons which apply in the case of the Mach 3 though the H2 did not enjoy the affection that was lavished, in an ambivalent way, on the 500. If owning a triple meant something of a love/hate relationship, then hate was the dominant factor where the H2 was concerned. Both bikes finally were killed off by American's increasingly stringent anti-pollution laws, which made life difficult for two-stroke fanciers, however artful the Injectolube or Autolube systems.

Suzuki 500 twin

Motorcyclists do not cover big mileages; and they change their bikes frequently. For these reasons it is not easy to establish the life expectancy of any particular model or, more interestingly, of its engine. BMW owners claim carefree miles running into tens of thousands. It is a brave man who argues with the committed owner of 'probably the best motorcycle made'. Note the insertion of 'probably' into the legend, to redeem what otherwise might come over as tub-thumping. 'Probably', by its very reasonableness, makes the message stronger. It is as if the boys from Berlin, calm and appraising, were willing to concede that somewhere out there, perhaps in the galaxy, a better motorcycle than the BMW might exist. But in the world as we know it? Hardly . . . The Japanese, by contrast, have not managed to project this aura of superiority, though their motorcycles, cheaper than many European equivalents on a straight comparison basis — infinitely cheaper if due account is given to the more complex technology embodied in the Japanese machines — obviously have a high-mileage potential. This is by way of introducing what might be thought of as an unlikely candidate for classic honours, the 500cc twin-cylinder two-stroke Suzuki, b.1968,d.(lingeringly) 1977. A ride-to-work machine for many people, on account of its low price, this Suzuki was fully a match in engine durability for any four-stroke twin. It was, also, a lot peppier. And much thirstier. Known at first as the Titan, then in the UK having its name changed to Cobra, the 500 Suzuki was brought to the market in face of the accumulated wisdom of a motorcycling public which declared that two-strokes worked as a 250cc twin (at 125cc per pot) and possibly as a 350, but were certain to vibrate and seize-up when enlarged to a full 250 x 2 = 500cc.

Long in the wheelbase, at 60 inches, low and more than a little uncertain in its handling, this not so humble device remained in production for almost a decade with no changes to speak of other than a mandatory annual rethink on tank colouring and, in 1971, a switch to disc braking at the front. Yet the Titan, later Cobra, finally Mk 3, did great things on the race-track in Britain and the USA. Eddie Crooks, a dealer in Lancashire, at one time saddled with Suzuki race sponsorship, years ago defended the

Suzuki's high-speed ways (someone had dared to suggest that top speed for the twin was nearer 100 mph than 120, as the advertisements claimed). Said Mr Crooks.

'120 indicated was commonplace on any Suzuki 500 I have ridden with low bars, and probably works out at a true 116 mph, which is the speed recorded by my production Suzuki ridden by the eventual winner, Frank Whiteway, in the 1970 TT. The Suzuki had previously won the 500 Production Class in the North-West 200, ridden this time by Stuart Graham. Preparation before the race consisted of a gearbox oil-change and a check on ignition timing. Everything else appeared to be A1. We did, however, treat it to a couple of new plugs before the TT, but never so much as lifted the cylinder heads. The same model was ridden in the Senior Manx Grand Prix by Les Trotter, who gained a Silver Replica, and before-race preparation on that occasion amounted to a set of brake linings and a new rear tyre. Total expenditure for the four races came to two gearbox oil changes, two plugs, one tyre and one pair of race brake linings. Frank Whiteway's win in the Production TT was probably one of the easiest rides he'd had, and he finished a comfortable 1½ minutes in front of the second and third men, on Daytona Triumphs. The machine was purchased from Suzuki GB Ltd, and was in every way an over-the-counter model, perfectly standard, other than for the usual fairing, tank, racing tyres, rims and seat. We even had air filters fitted. Various people have had ridiculous ideas about racing crankshafts, barrels and so on . . . all quite untrue!'

A milder classic than the Mach 3, Suzuki's 500 twin was durable, fast and in its day deservedly popular. The frame lacked a seat-to-gearbox tube, and hence some important rigidity, which made itself felt on occasion.

Bridgestone 350 GTR

Another early Japanese worthy of nomination as 'classic' is the Bridgestone 350 GTR. When he first saw one of these elegant two-strokes, at the 1967 Earls Court Show, Philip Vincent, designer of the world-famous vee-twin named after him, was moved to say:

'It is most attractive. Bridgestone use disc valves, like Kawasaki, and similarly enclose the carburettor, with air feed from a single air cleaner located high under the nose of the saddle. The high-performance 350 is called the GTR, and is claimed to develop 40 bhp at 7,500 rpm and to have a top speed of between 100 and 110 mph, to reach 60 mph in 5.2 seconds and to cover a standing quarter-mile in 13.7 seconds. With the machine weighing 363 lb, these performance figures seem to be compatible with 40 bhp, and it appears that this two-stroke twin would accelerate better than the four-stroke Honda 450 twin, handicapped by its considerably higher weight. The Bridgestone is a six-speeder, and a most novel and useful feature is the manner in which the kickstarter quadrant meshes with the primary gear train, enabling the engine to be started by lifting the clutch and while leaving a gear engaged. Very useful in thick traffic or on very steep slopes.'

I borrowed 'my' GTR from Dr M.R. Wigan (mentioned earlier) who was working at the Road Research Laboratory in Berkshire but chose, when I acknowledged his kindness in loaning me his treasured mile-eater, to be resident at 'Hertford College, Oxford'. Whether this was in tribute to Oxford, or was intended to turn aside possible reproaches from his employers at the RRL who might be incensed at publicity for a Japanese product, or was merely a crafty move in tax avoidance I was never to know. But I was grateful to Dr Wigan, then and on many future occasions, when his apparently limitless purse was able to satisfy his yearning for the very latest in motorcycling bijouterie which he would then, very decently, hand over to me for a while.

At the time, 1968, British manufacturers appeared in the main to have settled for 650cc and over. Only occasionally — nostalgically? — did they drop to 500cc. It was ironical that the Japanese had seen fit to move out of their up-to-250cc territory and on the way take over that once peculiarly British 'in between' category, the three-fifty. While the English were selling their bikes, quite successfully, on the North American market as 'he-man' irons, warts (vibration) and all — subtly implying that the Japs were a shade limp-wristed with their under-sized, under-powered, self-starting (ugh!) products — the Orientals were launching a counter-offensive by cramming in a few extra ccs, to end up with bikes having a top speed only a shade under a Bonneville's. The Bridgestone's all-aluminium engine was, at first sight, rather bulky. It was wide (18in) at the carburettor covers but not uncomfortably so over the footrests, which is what mattered. We made the measurement here about 14in average.

The inclined cylinders were made, as were the pistons, in silicon-alloy giving a constant expansion rate which was of benefit in providing for smaller than usual working clearances. In the Bridgestone design a chamber attached to the crankcase housed a disc geared to the crankshaft; the induction port was led into one side of this chamber and mixture from the carburettor was taken in a volume controlled by the position of the disc, which was cut away on its edge for the purpose. As the piston rose so the cutaway section of the disc came adjacent to the port's entry, and the mixture was inhaled; and then the disc moved round to block the port and the piston's down stroke. For the rest, the engine was almost conventional. There were ports to take the charge up into the cylinder, and one exhaust port at the front per cylinder.

The automatic oiling system was operated by an engine-driven pump mounted under the offside carburettor and controlled by throttle setting. Oil was fed to the outer main bearings from between the rotary disc and the mains, some part of this feed then going out to the disc. From the mains lubricant was taken to the hollow cranks, then into the needle-roller big-end bearings, thence to the cylinder. The four main bearings were particularly robust, with the two between the rods separated by an all-metal oil seal. Primary drive was by helical gears from the crankshaft, and the clutch was mounted away from the transmission, and thus could be of dry-plate construction, in racing style. Which meant that the clutch was practically faultless, with none of the tendency to dragging, or slipping, that bedevilled more usual oil-bath mechanisms. The six-speed gearbox, direct gear operating on fourth, ran mainly on ball bearings and the control for it was a lever on the left side, with neutral set above first. Though the Japanese supplied the bike with the

Bridgestone GTR350 had performance in the 1960s to match four-strokes 15 years on. Induction was by disc valve.

lever on the left, it could be swapped to the right — the then 'British' side — very easily, because the selector shaft emerged at either end of the transmission and was splined at both ends. And because the Japanese had thought everything through, there was correspondingly little trouble in changing over the brake to suit. The kickstarter action was taken direct to the crankshaft, as was noted by P.C. Vincent in his Earls Court walkabout. The benefits of this arrangement were negated, in a sense, by the ease with which one could find neutral, sometimes *very* elusive on other makes, merely by hooking up the lever as far as it would go.

In 1968 Japanese bikes seemed a shade undersized to the average Westerner — in physical bulk, that is; the engines tended to be small, and frame and fittings went the same way. But that could not be said of the Bridgestone, which had a strictly British-style wheelbase measurement of 55 inches and a dualseat over 30 inches high. One might grant classic status to the GTR solely on account of its seat, which was covered in cloth rather like the Bedford cord used in old American cars, at a time when all other manufacturers seemed satisfied with shiny plastic. Of course, it might have been a subtle move by Bridgestone to augment essential *grip* between rider and motorcycle during those talked-about 0 to 60 mph dashes. Certainly it was cool and resilient, narrow in the right places and adequately long for two. The frame was big, of twin-loop design, with a hefty no-flex mounting for the rear-fork pivot. The outer ends of the pivot were carried by the main-frame loops and the centre pivot housing was cut away and held by a vertical tube running from the top of the frame. Eminently practical, if old-fashioned, cycle-type adjusters were provided to push the wheel back, to take up chain slack, and of course the spindle nuts were split-pinned, in the usual Japanese manner. (In this area British manufacturers displayed more ingenuity and optimism than the Japanese with various eccentric cam adjustment methods and nut security *sans* pin.) The engine of the Bridgestone was held in the frame by one through bolt at the rear and one at the front, both in rubber to smooth insidious vibes, and by two vertical bolts from beneath the engine.

The rear-suspension units could be adjusted for 'ride', after the manner of the Velocette of the day, with the tops of the units moving back or forward, the variation being limited to two fixed positions, as opposed to the Velocette's infinite (within five inches) range. Dr Wigan, enthusiast that he was and even in those days carrying perhaps a shade more weight than was good for him, had changed the as-bought Japanese spring units for 90 lb Girlings, plus a British Avon tyre. On test rides I kept to Dr Wigan's recommended 5-star fuel — unnecessary, perhaps, in view of the 9.3:1 compression ratio and Bridge-

stone's talk about 'regular gasoline'. The sparking plugs were NGK B-8H which were not hard enough, according to the instruction book, but in practice gave satisfactory results. The wheels took 19 inch tyres, an inch or so bigger than average Japanese, and the brakes were of 7 inch diameter, the front with leading shoes. The Mikuni 26mm carburettors, concealed in light-alloy covers, had a large air-cleaner and the long, very efficient silencers were fitted with removable baffles. Unlike Honda's set up, which had two dials, as it were, under one glass, the Bridgestone's speedometer and rev-counter were side by side, with white needles and figures on black. The speedo was accurate, the rev-counter suffered from perceptible lag. A light came on to show when fifth gear was engaged, which was a good idea because of the smoothness of the engine, which could persuade you that fourth was top gear. The light showed also when you were in neutral. The tachometer was calibrated to 9,000 rpm, with a yellow band picking out the sector from 6,000 to 8,000 to indicate the high-torque range. Gear speed ranges, I found with too little regard for Dr Wigan's careful instructions, were up to 30 mph in first, 15 to 50 in second, 22 to 62 in third, 27 to 75 in fourth, 35 to 85 in fifth and 40 to 104 in top. The engine started easily, pulled smoothly all the way and developed plenty of power from 4,000 rpm, with the needle jetting into the yellow sector. The brakes were very good. Early miles on Dr Wigan's baby were devoted to exercising the gear lever and guessing which gear was engaged. When the fascination of this pastime faded it was simply a matter of pushing the lever up or down to cope with changes in road situation. One move, two — very rarely more — and the *right* gear would turn up. Top gear ratio was 5.64:1 and was not in any way an overdrive, for there was sufficient power to have it as the normal cruising-speed gear down to as low as 30 mph.

The Bridgestone in 1968 showed the sort of advances made by, say, Lamborghini in a more exalted sphere. A motorcyclist could take his Bridgestone among knowledgable motorists and be certain that the bike would excite interest. No middle-aged speedster would wind down his window and say — as he might, if the motorcyclist was on a 50 bhp four-stroke twin — 'Yes, I used to ride one of those before the war'. If you were a motorcyclist and you bought a Bridgestone, it would be a tribute to your educated taste. It was a classic.

Suzuki GT750

Another classic was the water-cooled Suzuki three, the GT750, which enjoyed only moderate success in its role of high-speed, effortless tourer on the road, but ended a brief life in some glory as progenitor of the TR750 racer which helped to make Barry Sheene unbeatable in F750 racing. The GT was Suzuki's entry into the so-called 'superbike' class created two years earlier by the CB750. Making it a three was not some misdirected effort at originality (vitiated in any case by the Kawasaki 500 triple) but showed good sense in permitting eventual use of components from the two-fifty single. Where Kawasaki, despite their success with air cooling, had been put to considerable trouble in cooling the middle cylinder and keeping overall width in bounds, Suzuki had leapt those obstacles by going for water-cooling, reaping the additional benefits of quietness and, with rubber mountings, vibration-free running. The crankshaft was carried in four bearings, with

the alternator at the righthand end and the contact-breaker at the other; primary drive was by helical gear taken from between the righthand and middle cylinders, with the electric starter mounted on the left. The five-speed gearbox had a take-off gear for the Posiforce lubrication system for the mains. Running on a (corrected) compression ratio of 6.7:1, this 70 x 64mm piston-ported three gave 67 bhp at 6,500 rpm and torque of 56.7 lb ft/5,500 rpm. Weight was around 550 lb.

When I first rode the GT750, I explained that the water-cooling gave extra mechanical hush, as the piston/bore clearances were set very close by virtue of more even temperatures. And then there was the muffling effect of the water-jacketing, and more consistent, possibly higher, power output than might be expected from an air-cooled two-stroke. So far as the GT750 was concerned, all this was largely theoretical, for the power produced was inferior to a Kawasaki's (even in 500 form), and the Suzuki's 67 horses were no match for the CB750's 67, either. The drawbacks in water-cooling

Suzuki GT750 watercooled three: well-mannered tourer that became a world-class racer.

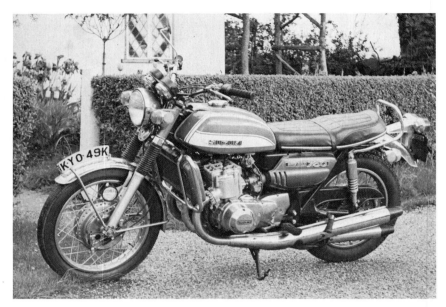

Country setting for a GT750 that shared a 'superbike' label with the Honda CB750 for a brief period around 1970.

The GT's three-cylinder engine was credited (by Suzuki) with power output identical to the CB750's. But on the road the four-stroke was noticeably faster in acceleration and top speed. The Suzuki, though, was smoother and possibly even more 'interesting'.

I enumerated as extra complication, weight and possibly some loss in looks. That Suzuki were particularly aware of the last-named was shown by vestigial finning on the highly polished water jacket, there presumably to maintain some visual link with mainstream (air-cooled) motorcycling. What seemed to be easily a 5-gallon tank turned out to hold about 3½, for at the nose it had to accommodate the header tank, accessible through a hinged flap, for the one-gallon cooling system which was controlled by an impeller in the bottom of the crank-

case. A thermostat restricted the quantity of water in circulation until a temperature of 82° was reached: a small electric fan mounted behind the radiator, barely visible and never audible because it was never in action, was set to engage when water temperature reached 106°C. In my time with the Suzuki the temperature needle never got on nodding terms with the mid-way mark on the dial, 86°C, and certainly never approached 106°C. But the designers presumably had to bear in mind Americans undertaking low-gear work in the Mojave desert as well as chilly Britons.

The three went very well. It was smooth, quiet mechanically and on the exhaust, which had the middle pipe branching into a silencer on each side, to make a more pleasing balance of two each side. The gearchange was delightful, roadholding was — well, very reasonable; and steering was weighty if steady. Acceleration was impressive, though hardly head-jerking, with waves of torque thrusting on to 105 mph top speed. When it was cruising for 100 miles on motorways at between 85 and 100 mph, which was illegal but not too nerve-wracking with the help of blur-free rear-view mirrors, the Suzuki gave a relaxed ride which was attributed to the lack of vibration from the engine. Some out of balance forces were seen to be at work, however, in the frenzied back and

forth motion of the forks at between 80 and 90 mph. In 100 miles a little over three gallons of 95 octane were used, working out to 31 mpg, which could not be construed as startling economy but stood up to figures returned by two-strokes of less than half the Suzuki's size. The brakes in those days had four leading shoes at the front, a single-leading design at the rear, and were not very good. Later the front drum, rather a good-looking set comprising two units from a 500 twin placed back to back, was replaced by twin discs.

For a long time I was undecided between buying a GT750 and Honda's four; finally I went for the Honda, and probably it was the better choice, though it was never as smooth as a Suzuki. (When the makers modified the Suzuki for F750 racing, power soared to well over 100 bhp, maximum speed to 170 mph and petrol consumption dropped to 15 mpg. It *must* have been a good engine.

above
On the production line at Hamamatsu. Visitors to Suzuki, or to any other Japanese plant, attracted little attention from assembly-line workers. Scarcely a head would be raised.

below
GT750 in its final form with twin front discs and sharper styling. A classic in any guise.

Marusho Lilac

Generalizations in motorcycling matters can be as fallacious as in any other sphere — more so, because motorcyclists, in anything to do with motorbikes, are blinkered to a degree, unwilling to concede that any make or model other than their own can have much to be said for it. They are forever ready to pass on legend as fact. Thus, on a 'national' rating, German motorcycles reputedly are well-made but rather dull; Italians have good design and engineering but bad finish; and Japanese bikes are cheaply made, reliable and imitative. So motorcyclists who compliment them-selves on their good taste in buying Italian 'thorough-bred' (the almost obligatory word) Moto Guzzi vee-twins, with the cylinders set across the frame, would not credit the

Marusho Lilac. The vee-twin was built in various sizes but the 300 was the main one for export to Europe.

Japanese with having more flair than the swarthy folk at Mandello del Lareo. But long before Guzzi's one-time industrial engine was trimmed and levered into a motorcycle frame, the classic Marusho Lilac 250 with transverse vee-twin engine, shaft drive and electric starting was on the roads of Japan and, in small numbers, Europe.

In the early 1960s the Marusho factory, founded in 1948, remained one of the smallest motorcycle plants in Japan, with an output of little more than 2,000 units a year. The Lilac was built in 125, 250 and 300cc versions, all with similar layout. The 250 was imported into Holland, and one or two were traded on an individual basis into the UK where one came to my attention.

Gerhard Klomps in Holland had ridden a Lilac and found it interesting. He says:

'For a European rider the Lilac had better overall dimensions than many other Japanese machines; certainly it suited me very well. I was extremely comfortable in the cushioned dualseat, knees tucked in against the humped, narrow, four-gallon tank in chromium and yellow. The duplex frame was black and seemed well designed, giving a taut feeling. Suspension was good, too . . . not too "soft". On Lilacs electric starting was introduced as early as 1954, and on this vee-twin was backed up, for emergencies, by a small kickstarter working transversely and fitted on the left side, just above the double gear change pedal. The gear change was of rotating type, such as the Honda had on its first appearance in Europe. To warn the rider about the danger of ending up in neutral or first gear, the makers had a warning light in the speedo face to say when third, or top, gears were engaged'.

Starting by press button, the engine had a nice 'four-stroke' burble . . . Klomps

Smooth, beautifully finished, the Lilac sold in only small numbers.

described it as 'a cross between Harley Davidson and BMW. The note was very quiet, as was the engine. When the throttle was opened, the twisting effect common to all machines having the crankshaft in line with the wheelbase was obvious but not to any worrying extent. Between 4,500 and 7,500 it ran as smoothly as an electric motor and without the slightest vibration'. Possibly Klomps was indulging in uncharacteristic exaggeration here, but it is clear that he was impressed. It peaked at 7,800 rpm and delivered 18.5 bhp. The makers claimed a top speed of 81 mph. Wurst-reared Klomps, in bulky riding gear, managed a true 72, sitting upright. The engine proved to be very flexible. A minimum speed of 15 mph could be maintained in top gear without transmission shock, and this was the more remarkable because shaft drive in those days had less capacity for shock-absorption than a chain. But spring absorbers were fitted in the single-plate clutch, and between gearbox and

drive shaft there was a flexible connection with metal bushes vulcanized into the material.

Top speed in second gear was 44, in third 55. The gear change was light and accurate, and almost noiseless if there was judicious slowing of the change. Fuel consumption worked out at about 73 mpg.

Beautifully designed, and a good-looker, the Lilac did not make it into the big time. To vindicate my earlier remarks about the uninformed prejudice that often governs the motorcyclist's choice, I should like to say there was no good reason for neglect of the Lilac. But it had crankshaft troubles, and the factory was slow in supplying modified components. The word spread (in Europe, at least) that Lilacs were troublesome. It was rumoured that the factory was to close. As far as big sales were concerned, that was the end. But the Lilac is a classic.

167

Today's Machines: A Market Guide

Note
The specifications given in this section have been derived from manufacturers' information and therefore differ in extent and units of measurement.

HONDA

Though some variations among the Honda range are not listed here, the chief new models introduced for 1981 are given pride of place. One of the most notable is the latest CBX-B, where the six-cylinder unit (in black) is offered in a package incorporating a full fairing as standard and new cycle parts using a version of Pro-Link rear suspension, first developed on the moto-crossers. Engine capacity remains 1,047cc but power characteristics are changed with new exhaust system, cams and timing to produce more mid-range torque (and a drop in peak output, to 100 bhp, in deference to German market requirements).

Honda CBX-6

Engine: Four stroke, six cylinder ohc, 1,047cc (64.5 x 53.4mm). Compression ratio, 9.3:1. Max power, 100ps/9,000 rpm. Max torque, 8.5kg/7,500rpm. Electric starter. Lubrication, wet sump, dual pump system with oil cooler. Carburettor, Keihin VB28 x 6. Five speed gearbox. Primary/secondary reduction: 1.000/2.269:1. Internal ratios: 2.439, 1.750, 1.391, 1.200 and 1.037:1 (top gear). Frame: Diamond. Telescopic front fork, swing arm rear. Tyres, 3.50 x 19 front, 130 x 90 rear. Castor angle, 62°30'. Trail, 120mm. Electrical equipment: 12v battery with 55w headlamp. Dimensions: Length, 2,325mm. Width, 730mm. Wheelbase, 1,535mm. Ground clearance, 155mm. Seat height, 810mm. Dry weight, 272kg. Fuel-tank capacity, 22 litres.

Honda CB900F-B

Honda's top-selling big bike remains the four-cylinder 16-valve CB900F; on both the 1981 B versions dual-piston disc brakes make an appearance. In addition, the front forks are linked by a hose to a single charging valve, for ease in setting the air preload. The F2-B has a red-and-white colour scheme and gold reversed Comstar wheels. The effect is rather good.

Engine: Four stroke, four cylinder, dohc, 901cc (64.5 x 69mm). Compression ratio, 8.8:1. Max power, 95ps/9,000rpm. Max torque, 7.9kg/8,000rpm. Starter motor.

Lubrication, forced, wet sump. Carburettor, Keihin VB32. Five speed gearbox. Primary/secondary reduction: 1.000/2.041:1. Internal ratios: 2.533, 1.789, 1.391. 1.160 and 1.000:1 (top gear). Clutch, wet multi-plate. Frame: Double cradle. Telescopic front fork, swing arm rear. Tyres, 3.25 x 19 front, 4 x 18 rear. Castor angle, 27°30'. Trail, 115mm. Electrical equipment: 12v battery with 50w headlamp. Dimensions: Length, 2,240mm. Width, 805mm. Wheelbase, 1,515mm. Ground clearance, 150mm. Seat height, 815mm. Dry weight, 234kg. Fuel-tank capacity, 20 litres.

Honda GL1100DX-B

Engine: Four stroke, watercooled, four cylinder ohc, 1,085cc (75 x 61.4mm). Compression ratio: 9.2:1. Max power, 83ps/7,500rpm. Max torque, 9.2kg/5,500rpm. Starter motor. Lubrication, forced, wet sump. Carburettor, constant velocity. Five speed gearbox. Primary reduction: 1.708. Internal ratios: 2.500, 1.667, 1.286, 1.065 and 0.909:1 (top gear). Clutch, wet multi-plate. Frame: Double cradle tubular steel. Telescopic front fork and swing arm rear. Tyres, 110/90 x 19 front, 130/90 x 17 rear. Castor angle, 60°50'. Trail, 134mm. Electrical equipment: 12v battery with 55w headlamp. Dimensions: Length, 2,355mm. Width, 825mm, Wheelbase, 1,605mm. Ground clearance, 155mm. Seat height, 795mm. Dry weight, 285kg. Fuel-tank capacity, 20 litres.

Honda CX500C-B

Engine: Four stroke, watercooled, two cylinder ohv, 496cc (78 x 52mm). Compression ratio, 10:1. Max power, 37kw/9,000rpm. Max torque, 43Nm/7,000rpm. Starter motor. Lubrication, forced, wet sump. Carburettor, constant vacuum piston valve. Five speed gearbox. Primary reduction: 2.242:1. Internal ratios: 2.733, 1.850, 1.416, 1.148 and 0.931:1 (top gear). Clutch, wet multi-plate. Frame: Diamond. Telescopic front fork, swing arm rear. Tyres, 3.50 x 19 front, 130/90 x 16 rear. Castor angle 26°45'. Trail, 105mm. Electrical equipment: 12v battery with 55w headlamp. Dimensions: Length, 2,240mm. Width, 875mm. Wheelbase, 1,455mm. Ground clearance, 145mm. Seat height, 790mm. Dry weight, 205kg. Fuel-tank capacity, 12 litres.

Honda CB400N-B

Engine: Four stroke, two cylinder ohc, 395cc (70.5 x 50.6mm). Compression ratio, 9.3:1. Max power, 43ps/9,500rpm. Max torque, 3.3kg/8,000rpm. Starter motor. Lubrication, forced, wet sump. Carburettor primary/secondary. Six speed gearbox. Primary reduction: 3.125:1. Internal ratios: 2.733, 1.947, 1.545, 1.280, 1.074 and 0.931:1 (top gear). Frame: Diamond. Telescopic front fork, swing arm rear. Tyres, 3.60 x 19 front, 4.10 x 19 rear. Castor angle, 63°. Trail, 100mm. Electrical equipment: 12v battery with 55w headlamp. Dimensions: Length, 2,165mm. Width, 740mm. Wheelbase, 1,390mm. Ground clearance, 165mm. Seat height, 795mm. Dry weight, 173kg. Fuel-tank capacity, 14 litres.

Honda XL250S-B

The four-stroke trail bikes continue, with small changes. The XL250S-B and the 185 have new instrument panels, with an extra rev-counter for the 250. A windshield protects the front of the instruments and an extra chain guide has been fitted to both bikes, together with fork gaiters for all except the 100, a beginner's machine first shown in Europe at Bristol in December 1980.

Engine: Four stroke, single cylinder ohc, 249cc (74 x 57.8mm). Compression ratio, 9.1:1. Max power, 20.4hp/7,500rpm. Max torque, 2.04kg/6,000rpm. Kick starter with automatic compression release. Lubrication, forced, wet sump. Carburettor, piston valve type. Five speed gearbox. Primary reduction: 2.399:1. Internal ratios: 2.800, 1.850, 1.375, 1.111 and 0.900:1 (top gear). Frame: Diamond. Telescopic front fork, swing arm rear. Tyres, 300 x 23 front, 460 x 18 rear. Castor angle, 61°30'. Trail, 138mm. Electrical equipment: 6v battery with 35w headlamp. Dimensions: Length, 2,175mm. Width, 875mm. Wheelbase, 1,390mm. Ground clearance, 260mm. Seat height, 850mm. Dry weight, 123kg. Fuel-tank capacity, 9.5 litres.

Honda CB250NDX-B

Engine: Four stroke, two cylinder ohc, 249cc (62 x 41.4mm). Max power, 20hp/10,000rpm. Max torque, 19.6kg/8,500rpm. Compression ratio, 9.4:1. Lubrication, forced, wet sump. Carburettor, primary/secondary. Six speed gearbox. Starter motor. Frame: Diamond type. Telescopic front fork, rear swing arm. Tyres, 3.60 x 19 front, 4.10 x 18 rear. Trail, 100mm. Castor angle, 27°. Electrical equipment: 12v battery with 40w headlamp. Dimensions: Length, 2,115mm. Width, 730mm. Height, 1,105mm. Wheelbase, 1,395mm. Ground clearance, 165mm. Dry weight, 169kg. Fuel-tank capacity, 14 litres.

Honda XL185S-B

Engine: Four stroke, single cylinder ohc, 180cc (63 x 57.8mm). Compression ratio, 9.2:1. Max power, 16ps/8,000rpm. Max torque, 1.5kg/6,500rpm. Kick starter. Lubrication, forced, wet sump. Carburettor, piston-valve type. Five speed gearbox. Primary reduction: 3.333:1. Internal ratios: 2.759, 1.722, 1.272, 1.000 and 0.777:1 (top gear). Clutch, wet multi-plate. Frame: Diamond. Telescopic front fork, swing arm rear. Tyres, 2.75 x 21 front, 4.10 x 18 rear. Castor angle, 37°50'. Trail, 122mm. Electrical equipment: 6v battery with 35w headlamp. Dimensions: Length, 2,130mm. Width, 855mm. Wheelbase, 1,310mm. Ground clearance, 270mm. Seat height, 820mm. Dry weight, 107.5kg. Fuel-tank capacity, 7 litres.

Honda XL125S-B

Engine: Four stroke, single cylinder ohc, 124cc (56.5 x 49.5mm). Max power, 13ps/9,500rpm. Max torque, 1.05kg/8,000rpm. Compression ratio, 9.4:1. Lubrication, forced, wet sump. Carburettor, piston valve type. Six speed gearbox. Primary reduction: 3.333:1. Internal ratios: 3.083, 1.941, 1.400, 1.130, 0.925 and 0.785:1 (top gear). Kick starter. Frame: Diamond. Telescopic front fork, rear swing arm. Tyres, 2.75 x 21 front, 4.10 x 18 rear. Trail, 122mm. Castor angle, 27°50'. Electrical equipment: 6v battery, 35w headlamp. Dimensions: Length, 2,170mm. Width, 840mm. Height, 1,110mm. Wheelbase, 1,310mm. Ground clearance, 270mm. Dry weight, 106kg. Fuel-tank capacity, 7 litres.

Honda CD200T-B

Engine: Four stroke, single cylinder, 194cc (53 x 44mm). Compression ratio, 8.8:1. Max power, 16.1hp/8,500rpm. Max torque, 1.52kg/6,500rpm. CDI. Electric/kick starter. Lubrication, forced, wet sump. Carburettor, piston valve type. Four speed gearbox.

Primary reduction: 3.632:1. Internal ratios: 2.846, 1.778, 1.273, 1.000:1 (top gear). Clutch, wet multi-plate. Frame: Diamond. Telescopic fork front, swing arm rear. Tyres, 3 x 17 front and rear. Castor angle, 62° 40¹. Trail, 89mm. Brakes, drum front and rear. Electrical equipment: 12v battery with 45w headlight. Dimensions: Length, 1,970mm. Width, 685mm. Wheelbase, 1,280mm. Ground clearance, 150mm. Seat height, 745mm. Dry weight, 128.5kg. Fuel-tank capacity, 11 litres.

Honda CG125K-B

Engine: Four stroke, single cylinder, 124cc (56.5 x 49.5mm). Compression ratio, 9:1. Max power, 11hp/9,000rpm. Max torque, 0.94kg/7,500rpm. CDI. Kick starter. Lubrication, forced, wet sump. Carburettor, piston valve type. Five speed gearbox. Primary reduction: 4.055:1. Internal ratios: 2.769, 1.882, 1.450, 1.173 and 1.000:1 (top gear). Clutch, wet multi-plate. Frame: Diamond. Front telescopic fork, rear swing arm. Tyres, 2.50 x 18 front, 3 x 187 rear. Castor angle, 64°. Trail, 80mm. Brakes, drum front and rear. Electrical equipment: 6v battery with 25w headlight. Dimensions: Length, 1,840mm. Width, 735mm. Wheelbase, 1,200mm. Ground clearance, 135mm. Seat height, 755mm. Dry weight, 95kg. Fuel-tank capacity, 10 litres.

Honda CB125T-B

Engine: Four stroke, two cylinder, 124cc (44 x 41mm). Compression ratio, 9.4:1. Max power, 16.5hp/11,500rpm. Max torque, 1.04kg/10,500rpm. CDI. Kick starter. Lubrication, forced, wet sump. Carburettor, piston valve type. Five speed gearbox. Primary reduction: 3.833:1. Internal ratios: 2.769, 1.882, 1.450, 1.217 and 1.083:1 (top gear). Clutch, wet multi-plate. Frame: Diamond. Front telescopic fork, rear swing arm. Tyres, 2.75 x 18 front and 3 x 18 rear. Castor angle, 63°. Trail, 90mm. Brakes, front drum, rear disc. Electrical equipment: 6v

battery with 35w headlight. Dimensions: Length, 1,980mm. Width, 680mm. Wheelbase, 1,275mm. Ground clearance, 160mm. Seat height, 770mm. Dry weight, 114kg. Fuel-tank capacity, 11.5 litres.

Honda XL100S-B

Engine: Four stroke, single cylinder ohc, 99.2cc (53 x 54mm). Max power, 7.5ps/9,500rpm. Max torque, 0.7kg/8,000rpm. Compression ratio, 9.4:1. Lubrication, forced, wet sump. Carburettor, piston valve type. Five speed gearbox. Primary reduction: 4.437:1. Internal ratios: 3.083, 1.882, 1.400, 1.130 and 0.923:1 (top gear). Kick starter. Frame: Diamond. Telescopic front fork, rear swing arm. Tyres, 2.50 x 19 front, 3 x 16 rear. Trail, 108mm. Castor angle, 29°. Electrical equipment: 6v battery with 35w headlamp. Dimensions: Length, 1,920mm. Width, 805mm. Height, 1,060mm. Wheelbase, 1,225mm. Ground clearance, 255mm. Dry weight, 80kg. Fuel-tank capacity, 4.5 litres.

Honda MB50

Engine: Two stroke, single cylinder, 3.0 cu.in. Compression ratio, 6.4:1. Max power 2.55 ps/6,000rpm. Max torque 0.41kg/4,500rpm. CDI. Kick starter. Lubrication, 2 stroke oil injection. Carburettor, piston valve. Five speed gearbox. Primary reduction: 4.117:1. Internal ratios: 3.038, 1.882, 1.400, 1.130 and 0.960:1 (top gear). Clutch, wet multi-plate. Frame: Cross line backbone. Front telescopic fork, rear swing arm. Tyres, 2.50 x 18 front and rear. Trail, 2.8 in. Brakes, front disc, rear drum. Electrical equipment: 5v battery with 25w headlight. Dimensions: Length, 74 in. Width, 25.8 in. Height, 45.7 in. Wheelbase, 48.2 in. Seat height, 29.5 in. Ground clearance, 6.3 in. Dry weight, 81 kg. Fuel-tank capacity, 9 litres.

Honda NX50M-B Caren

Engine: Two stroke, single cylinder, 49cc (40 x 39.3mm). Compression ratio, 7.3:1. Max power, 3.1hp/5,500rpm. Max torque, 0.43kg/4,500rpm. CDI. Electric/kick starter. Lubrication, forced wet sump. Carburettor, piston valve type. Primary reduction v-belt. Final reduction, 10.244:1. Automatic centrifugal clutch, dry type. Frame: Backbone type. Front telescopic, rear swing arm. Tyres, 2.75 x 10, front and rear. Castor angle, 25°30. Trail, 65mm. Brakes, front and rear drum, 95mm. Electrical equipment: 12v battery with 25w headlight. Dimensions: Length, 1,568mm. Width, 620mm. Wheelbase, 1,095mm. Ground clearance, 110mm. Seat height, 715mm. Dry weight, 57kg. Fuel-tank capacity, 3 litres.

Honda C50L

Engine: Four stroke, single cylinder, 49cc (39 x 41.4mm). Compression ratio, 9.5:1. Lubrication, forced, wet sump. Carburettor, piston valve type. Three speed gearbox. Frame: Backbone type, bottom-link front fork, rear swing arm. Tyres, 2.25 x 17 front and rear. Trail, 75mm. Castor angle, 26°30. Electrical equipment: 6v/4ah battery with 25w headlamp. Dimensions: Length, 1,785mm. Width, 640mm. Height, 975mm. Wheelbase, 1,175mm. Seat height, 770mm. Dry weight, 71kg. Fuel-tank capacity, 3 litres.

Honda NC50K

Engine: Two stroke, single cylinder, 49cc (40 x 39.6mm). Compression ratio, 6.7:1. CDI. Kick starter. Frame: Backbone design. Front telescopic fork, rear swing arm. Tyres, 2.25 x 14 front and rear. Dimensions: Length, 530mm, Width, 605mm, Height, 975mm, Wheelbase, 1,019mm, Ground clearance, 120mm. Dry weight, 52kg. Fuel-tank capacity, 2 litres.

Honda ATC185 (provisional spec.)

Honda's conception of the three-wheeled work horse, the All Terrain Cycle, is the ATC 185S-B, which is powered by a sohc single with automatic clutch and five foot-operated gears. The frame is a double cradle, and there is a large bash plate under the rear axle, which pivots for chain ('o' ring type) tensioning. The throttle is operated by right thumb. All brakes are cable operated, and starting is by hand pull. High mounting of air filter and exhaust system allows Honda to claim that the ATC will ford 12 in of water with impunity.

Engine: Four stroke, ohc, 180cc (63 x 57.8mm). Compression ratio, 8:1. Max power, 13ps/7,000rpm. Max torque, 1.38kg/5,500rpm. Kick starter. Clutch, automatic centrifugal. Dimensions: Length, 1,710mm. Width, 1,000mm. Wheelbase, 1,115mm. Ground clearance, 110mm. Overall height, 950mm. Dry weight, 127kg.

KAWASAKI

Kawasaki Z1100-B1

The Z1100GP, a new-for-1981 four, has electronic fuel injection (to be found, among production bikes only on Kawasakis). This, apart from its extra efficiency compared with ordinary carburettors, is said to be easier to maintain. Well, perhaps . . . Top speed is around 132 mph. The engine finish is black, which the best brains in the business apparently now considers essential in promoting the go-faster image; in an odd way it is akin to the thinking behind making 'professional' cameras all-black, as opposed to aluminium finish for the punter's version. Altogether, it is a good-looking bike having perhaps the best handling of any of the big Kawasakis.

Engine: Four stroke, four cylinder dohc, 1,089cc (72.5 x 66mm). Compression ratio, 8.9:1. Max power, 108hp/8,500rpm. Max torque, 9.8kg/7,000rpm. CDI. Electric starter. Lubrication, forced, wet sump with oil cooler. Electronic fuel injection. Valve timing: Inlet, open 35°BTDC, close 65°ABDC. Exhaust, open 68°BBDC, close 32°ATDC. Five speed gearbox. Primary reduction: 1.732:1. Internal ratios: 2.642, 1.833, 1.428, 1.174 and 1.040:1 (top gear). Clutch, wet multi-plate. Frame: Tubular double cradle. Front air adjustable telescopic fork, rear swing arm. Tyres, 3.25V x 19 front, 4.25V x 18 rear. Castor angle, 29°. Trail, 120mm. Brakes, front dual discs, 236mm, rear disc, 236mm. Electrical equipment: 12v/16ah battery with 55w headlight. Dimensions: Length, 2,265mm. Width, 820mm. Wheelbase, 1,540mm. Ground clearance, 145mm. Seat height, 805mm. Dry weight, 237.5kg. Fuel-tank capacity, 21.4 litres.

Kawasaki Z1300-A3

Still the biggest, if not the fastest, Kawasaki . . . because even 120 bhp can be spread thin when heaving close to half a ton into the wide blue yonder. Nevertheless, the Z1300 is a wonder to behold; and yes, to ride . . .

Engine: Four stroke, water-cooled, six cylinder dohc, 1,286cc (62 x 71mm). Compression ratio, 9.9:1. Max power, 120hp/8,000rpm. Max torque, 11.8kg/6,500rpm. CDI. Electric starter. Lubrication, forced, wet sump. Carburettor, Mikuni BSW32 x 3. Valve timing: Inlet, open 20°BTDC, close 70°ABDC. Exhaust, open 70°BBDC, close 30°ATDC. Five speed gearbox. Primary reduction: 1.841:1. Internal ratios: 2.294, 1.666, 1.280, 1.074 and 0.931:1 (top gear). Clutch, wet multi-plate. Frame: Front air adjustable telescopic fork, rear air adjustable swing arm. Tyres, 110/90V x 18 front, 130/90V x 17 rear. Castor angle, 28°. Trail, 100mm. Brakes, front dual discs, 260mm, rear disc, 250mm. Electrical equipment: 12v/20ah battery with 55w headlight. Dimensions: Length, 2,335mm. Width, 840mm. Wheelbase, 1,580mm. Ground clearance, 150mm. Seat height, 820mm. Dry weight, 296kg. Fuel-tank capacity, 27 litres.

Kawasaki Z1100-A1

Shaft drive and refined suspension: the A1 1100 has a leading-axle, air-adjustable front fork with an equalizing tube to prevent uneven pressure. At the rear an air-adjustable system is similarly equipped, plus a four-position damping control.

Engine: Four stroke, four cylinder dohc, 1089cc (72.5 x 66mm). Compression ratio, 8.9:1. Max power, 100hp/8,000rpm. Max torque, 9.8kg/6,500rpm. CDI. Electric starter. Lubrication, forced, wet sump. Carburettor, Mikuni BS34 x 4. Valve timing: Inlet, open 30°BTDC, close 60°ABDC. Exhaust, open 63°BBDC, close 27°ATDC. Five speed gearbox. Primary reduction: 1.732:1. Internal ratios: 2.642, 1.833, 1.428, 1.174 and 1.040:1 (top gear). Clutch, wet multi-plate. Frame: Tubular double cradle. Front telescopic fork, rear swing arm. Tyres, 3.50V x 19 front, 130/90V x 16 rear. Castor angle, 29°. Trail, 125mm. Brakes, front dual discs, 236mm, rear disc, 236mm. Electrical equipment: 12v/18ah battery with 55w headlight. Dimensions: Length, 2,310mm. Width, 890mm. Wheelbase, 1,545mm. Ground clearance, 125mm. Seat height, 790mm. Dry weight, 246kg. Fuel-tank capacity, 21.4 litres.

Kawasaki Z1000-J1

The standard big Kawasaki drops to 998cc, gains a few horses, loses a little weight. The front fork is air-adjustable.

Engine: Four stroke, four cylinder dohc, 998cc (69.4 x 66mm). Compression ratio, 9.2:1. Max power, 102hp/8,500rpm. Max torque, 9.3kg/7,000rpm. CDI, Electric starter. Lubrication, forced, wet sump. Carburettor, Mikuni BS34 x 4. Valve timing: Inlet, open 35°BTDC, close 65°ABDC. Exhaust, open 68°BBDC, close 32°ATDC. Five speed gearbox. Primary reduction: 1.732:1. Internal ratios: 2.642, 1.833, 1.428, 1.174 and 1.040:1 (top gear). Clutch, wet multi-plate. Frame: Tubular, double cradle. Front telescopic fork, rear swing arm. Tyres, 3.25V x 19 front, 4.25V x 18 rear. Castor angle, 27.5°. Trail, 99mm. Brakes, front dual discs, 236mm, rear disc, 236mm. Electrical equipment: 12v/18ah battery with 55w headlight. Dimensions: Length, 2,265mm. Width, 820mm. Wheelbase, 1,520mm. Ground clearance, 140mm. Seat height, 805mm. Dry weight, 230kg. Fuel-tank capacity, 21.4 litres.

Kawasaki Z1000LTD-K1

Biggest of the LTD range, the 1000 is aimed at one type of rider: the man who wants plenty of power allied to American styling. It is, in fact, very comfortable, thanks to the 'dual-density' — and dual-height — seat, and suspension with a wide range of adjustment.

Engine: Four stroke, four cylinder dohc, 998cc (69.4 x 66mm). Compression ratio, 9.2:1. Max power, 92hp/8,000rpm. Max torque, 8.7kg/7,000rpm. CDI. Electric starter. Lubrication, forced, wet sump. Carburettor, Mikuni BS34 x 4. Valve timing: Inlet, open 30°BTDC, close 60°ABDC. Exhaust, open 63°BBDC, close 27°ATDC. Five speed gear-box. Primary reduction: 1.732:1. Internal ratios: 2.642, 1.833, 1.428, 1.174 and 1.040:1 (top gear). Clutch, wet multi-plate. Frame: Tubular, double cradle. Front telescopic fork, rear swing arm. Tyres, 3.25V x 19 front, 130/90V x 16 rear. Castor angle, 29°. Trail, 107mm. Brakes, front dual discs, 236mm, rear disc, 236mm. Electrical equipment: 12v/18ah battery with 55w headlight. Dimensions: Length, 2,293mm. Width, 820mm. Wheelbase, 1,535mm. Ground clearance, 130mm. Seat height, 785mm. Dry weight, 234kg. Fuel-tank capacity, 15 litres.

Kawasaki Z750-L1

In the 1000 mould, but lighter: that's the 750-LI.

Engine: Four stroke, four cylinder dohc, 738cc (66 x 54mm). Compression ratio, 9:1. Max power, 74hp/9,000rpm. Max torque, 6.4kg/7,500rpm. CDI. Electric starter. Lubrication, forced, wet sump. Carburettor, Keihin CV34 x 4. Valve timing: Inlet, open 30°BTDC, close 60°ABDC. Exhaust, open 60°BBDC, close 30°ATDC. Five speed gearbox. Primary reduction: 2.550:1. Internal ratios: 2.333, 1.631, 1.272, 1.040 and 0.875:1 (top gear). Clutch, wet multi-plate. Frame: Tubular, double cradle. Front telescopic fork, rear swing arm. Tyres, 3.25H x 19 front, 4.00H x 18 rear. Castor angle, 27°. Trail, 108mm. Brakes, front dual discs, 226mm, rear disc, 226mm. Electrical equipment: 12v/12ah battery with 55w headlight. Dimensions: Length, 2,190mm. Width, 780mm. Wheelbase, 1,420mm. Ground clearance, 150mm. Seat height, 810mm. Dry weight, 211kg. Fuel-tank capacity, 21.7 litres.

Kawasaki Z750LTD-H2

Engine: Four stroke, four cylinder dohc, 738cc (66 x 54mm). Compression ratio, 9:1. Max power, 74hp/9,000rpm. Max torque, 6.4kg/7,500rpm. CDI. Electric starter. Lubrication, forced, wet sump. Carburettor, Keihin CV34 x 4. Valve timing: Inlet, open 30°BTDC, close 60°ABDC. Exhaust, open 60°BBDC, close 30°ATDC. Five speed gearbox. Primary reduction: 2.550:1. Internal ratios: 2.333, 1.631, 1.272, 1.040 and 0.875:1 (top gear). Clutch, wet multi-plate. Frame: Tubular, double cradle. Front telescopic fork, rear swing arm. Tyres, 3.25H x 19 front, 130/90 x 16 rear. Castor angle, 30°. Trail, 121mm. Brakes, front dual discs, 226mm, rear disc, 226mm. Electrical equipment: 12v/12ah battery with 55w headlight. Dimensions: Length, 2,195mm. Width, 810mm. Wheelbase, 1,450mm. Ground clearance, 155mm. Seat height, 770mm. Dry weight, 211.3kg. Fuel-tank capacity, 12.4 litres.

Kawasaki Z650SR-D4

Engine: Four stroke, four cylinder dohc, 652cc (62 x 54mm). Compression ratio, 9.5:1. Max power, 82hp/8,500rpm. Max torque, 5.8kg/7,000rpm. CDI. Electric starter. Lubrication, forced, wet sump. Carburettor, Mikuni VM24SS x 4. Valve timing: Inlet, open 22°BTDC, close 52°ABDC. Exhaust, open 60°BBDC, close 20°ATDC. Five speed gearbox. Primary reduction: 2.550:1. Internal ratios: 2.333, 1.631, 1.272, 1.040 and 0.888:1 (top gear). Clutch, wet multi-plate. Frame: Tubular, double cradle. Front air adjustable telescopic fork, rear swing arm. Tyres, 3.25H x 19 front, 130/90 x 16 rear. Castor angle, 27.5°. Trail, 113mm. Brakes, front dual discs, 226mm, rear disc, 226mm. Electrical equipment: 12v/10ah battery with 55w headlight. Dimensions: Length, 2,185mm. Width, 830mm. Wheelbase, 1,440mm. Ground clearance, 155mm. Seat height, 795mm. Dry weight, 214.3kg. Fuel-tank capacity, 14 litres.

Kawasaki Z650-F2

Engine: Four stroke, four cylinder dohc, 652cc (62 x 54mm). Compression ratio, 9.5:1. Max power, 64hp/8,500rpm. Max torque, 5.8kg/7,000rpm. CDI. Electric starter. Lubrication, forced, wet sump. Carburettor, Mikuni VM24SS x 4. Valve timing: Inlet, open 22°BTDC, close 52°ABDC. Exhaust, open 60°BBDC, close, 22°ATDC. Five speed gearbox. Primary reduction: 2.550:1. Internal ratios: 2.333, 1.631, 1.272, 1.040 and 0.888:1 (top gear). Clutch, wet multi-plate. Frame: Tubular, double cradle. Front air adjustable telescopic fork, rear swing arm. Tyres, 3.25H x 19 front, 4.00H x 18 rear. Castor angle, 27°. Trail, 108mm. Brakes, front dual discs, 226mm, rear drum, 180 x 40mm. Electrical equipment: 12v/10ah battery with 55w headlight. Dimensions: Length, 2,220mm. Width, 775mm. Wheelbase, 1,440mm. Ground clearance, 152mm. Seat height, 820mm. Dry weight, 209kg. Fuel-tank capacity, 16.8 litres.

Kawasaki Z550-D1

Kawasaki got all their sums right with the Z550GP. It is light and agile, has plenty of power, and good looks. People with experience of the latest European machinery compare the 550 favourably with the best of the Ducatis on handling . . . and you can't say fairer than that.

Engine: Four stroke, four cylinder dohc, 553cc (58.0 x 52.4mm). Compression ratio, 10:1. Max power, 58hp/9,000rpm. Max torque, 4.9kg/8,000rpm. CDI. Electric starter. Lubrication, forced, wet sump, with oil cooler. Carburettor, TK K22P x 4. Valve timing: Inlet, open 31°BTDC, Close 59°ABDC. Exhaust, open 59°BBDC, close 31°ATDC. Six speed gearbox. Primary reduction: 2.935:1. Internal ratios: 2.571, 1.777, 1.380, 1.125, 0.961 and 0.851:1 (top gear). Clutch, wet multi-plate. Frame: Tubular, double cradle. Front air adjustable telescopic fork, rear swing arm. Tyres, 3.25H x 19 front, 3.75H x 18 rear. Castor angle, 26°. Trail,

98mm. Brakes, front dual discs, 226mm, rear disc, 226mm. Electrical equipment: 12v/12ah battery with 55w headlight. Dimensions: Length, 2,150mm. Width, 740mm. Wheelbase, 1,400mm. Ground clearance, 145mm. Seat height, 805mm. Dry weight, 199.5kg. Fuel-tank capacity, 15 litres.

Kawasaki Z550-A2

Engine: Four stroke, four cylinder dohc, 553cc (58 x 52.4mm). Compression ratio, 9.5:1. Max power, 54hp/8,500rpm. Max torque, 4.9kg/7,000rpm. CDI. Electric starter. Lubrication, forced, wet sump. Carburettor, TK K22P x 4. Valve timing: Inlet, open 20°BTDC, close 48°ABDC. Exhaust, open 48°BBDC, close 20°ATDC. Six speed gearbox. Primary reduction: 2.935:1. Internal ratios: 2.571, 1.777, 1.380, 1.125, 0.961 and 0.851:1 (top gear). Clutch, wet multi-plate. Frame: Tubular, double cradle. Front air adjustable telescopic fork, rear swing arm. Tyres, 3.25H x 19 front, 3.75H x 18 rear. Castor angle, 26°. Trail, 98mm. Brakes, front dual disc, 226mm, rear drum, 180mm. Electrical equipment: 12v/12ah battery with 55w headlight. Dimensions: Length, 1,395mm. Width, 740mm. Wheelbase, 1,395mm. Ground clearance, 145mm. Seat height, 805mm. Dry weight, 191kg. Fuel-tank capacity, 15 litres.

Kawasaki Z550LTD-C2

Engine: Four stroke, four cylinder dohc, 553cc (58 x 52.4mm). Compression ratio, 9.5:1. Max power, 54hp/8,500rpm. Max torque, 4.9kg/7,000rpm. CDI. Electric starter. Lubrication, forced, wet sump. Carburettor, TK K22P x 4. Valve timing: Inlet, open 20°BTDC, close 48°ABDC. Exhaust, open 48°BBDC, close 20°ATDC. Six speed gearbox. Primary reduction: 2.935:1. Internal ratios: 2.571, 1.777, 1.380, 1.125, 0.961 and 0.851:1 (top gear). Clutch, wet multi-plate. Frame: Tubular, double cradle. Front air

173

adjustable telescopic fork, rear swing arm. Tyres, 3.25 x 19 front, 130/90 x 16 rear. Castor angle 27.5°. Trail, 110mm. Brakes, front dual discs, 226mm, rear drum, 180mm. Electrical equipment: 12v/12ah battery with 55w headlight. Dimensions: Length, 2,190mm. Width, 850mm. Wheelbase, 1,420mm. Seat height, 770mm. Dry weight, 198kg. Fuel-tank capacity, 12.4 litres.

Kawasaki Z440-A2

Engine: Four stroke, two cylinder sohc, 443cc (67.5 x 62mm). Compression ratio, 9.2:1. Max power, 40hp/8,500rpm. Max torque, 3.6kg/7,000rpn. CDI. Electric starter. Lubrication, wet sump. Carburettor, CV36 x 2. Valve timing: Inlet, open 27°BTDC, close 73°ABDC. Exhaust, open 70°BBDC, close 30°ATDC. Six speed gearbox. Primary reduction: 2.434:1.

Internal ratios: 2.538, 1.750, 1.315, 1.095, 0.956 and 0.875:1 (top gear). Clutch, wet multi-plate. Frame: Tubular, double cradle. Front telescopic fork, rear swing arm. Tyres, 3.25S x 19 front, 130 x 16 rear. Castor angle, 27°. Trail, 112mm. Brakes, front single disc, 230mm, rear drum, 160mm. Electrical equipment: 12v/12ah battery with 35/35w headlight. Dimensions: Length, 2,120mm. Width, 810mm. Wheelbase, 1,390mm. Ground clearance, 140mm. Seat height, 740mm. Dry weight, 170kg. Fuel-tank capacity, 12 litres.

Kawasaki Z400-J2

Castigated as dull by one of the UK motorcycle papers, the Z400 represents Kawasaki's attempt to woo the sort of people who were happy with Honda's CB400F four. For the record: it isn't dull, but it *is* terribly efficient. Perhaps efficiency equals dullness?

Engine: Four stroke, four cylinder dohc, 399cc (52 x 47mm). Compression ratio, 9.5:1. Max power, 43hp/9,500rpm. Max torque, 3.5kg/7,500rpm. CDI. Electric starter. Carburettor, TK K21P x 4. Valve timing: Inlet, open 33°BTDC, close 41°ABDC. Exhaust, open 51°BBDC, close 23°ATDC. Six speed gearbox. Primary reduction: 3.277:1. Internal ratios: 2.571, 1.777, 1.380, 1.125, 0.961 and 0.851:1 (top gear). Clutch, wet multi-plate. Frame: Tubular, double cradle. Front air adjustable telescopic fork, rear swing arm. Tyres, 3.25S x 19 front, 3.75S x 18 rear. Castor angle, 26°. Trail, 98mm. Brakes, front dual discs, 226mm, rear drum, 180mm. Electrical equipment: 12v/12ah battery with 60/55w headlight. Dimensions: Length, 2,150mm. Width, 740mm. Wheelbase, 1,395mm. Ground clearance, 145mm. Seat height, 805mm. Dry weight, 191kg. Fuel-tank capacity, 15 litres.

Kawasaki Z250-A3

Engine: Four stroke, two cylinder sohc, 248cc (55 x 52.4mm). Compression ratio, 9.5:1. Max power, 27hp/10,000rpm. Max torque, 2.1kg/8,000rpm. CDI. Electric starter. Lubrication, forced, wet sump. Carburettor, Keihin CV32 x 2. Valve timing: Inlet, open 21°BTDC, close 59°ABDC. Exhaust, open 61°BBDC, close 19°ATDC. Six speed gearbox. Primary reduction: 3.736:1. Internal ratios: 2.600, 1.789, 1.409, 1.160, 1.000 and 0.892:1 (top gear). Clutch, wet multi-plate. Frame: Tubular, single cradle. Front telescopic fork, rear swing arm. Tyres, 3.00S x 18 front, 3.50S x 18 rear. Castor angle, 27°. Trail, 100mm. Brakes, front single disc, 226mm, rear drum 218mm. Electrical equipment: 12v/10ah battery with 35w headlight. Dimensions: Length, 2,015mm. Width, 740mm. Wheelbase, 1,340mm. Ground clearance, 140mm. Seat height, 805mm. Dry weight, 153kg. Fuel-tank capacity 13.6 litres.

Kawasaki KL250-A4

Engine: Four stroke, single cylinder sohc, 246cc (70 x 64mm). Compression ratio, 8.9:1. Max power, 21hp/8,500rpm. Max torque, 2.0kg/6,500rpm. CDI. Kick starter. Lubrication, forced, wet sump. Carburettor, Mikuni BS34. Valve timing: Inlet, open 27°BTDC, close 65°ABDC. Exhaust, open 62°BBDC, close 30°ATDC. Five speed gearbox. Primary reduction: 3.285:1. Internal ratios: 2.636, 1.733, 1.300, 1.050 and 0.875:1 (top gear). Clutch, wet multi-plate. Frame: Tubular, single cradle. Front telescopic fork, rear swing arm. Tyres, 3.00 x 21 front, 4.60 x 17 rear. Castor angle, 30.5°. Trail 131mm. Brakes, drum, 120mm, front and rear. Electrical equipment: 6v/6ah battery with 35w headlight. Dimensions: Length, 2,240mm. Width, 885mm. Wheelbase, 1,415mm. Ground clearance, 240mm. Seat height, 855mm. Dry weight, 118kg. Fuel-tank capacity, 9.8 litres.

Kawasaki Z250-C2

Fifteen years after European (mainly British) factories gave up 250 singles, and just about everything else, in face of the onslaught of Japanese multis, the Japanese have, apparently, discovered the sales potential of a light, economical, uncomplicated, lively single-cylinder motorcycle of around 250cc. Kawasaki's version is one of the best.

Engine: Four stroke, single cylinder sohc, 246cc (70 x 64mm). Compression ratio, 8.9:1. Max power, 19hp/8,000rpm. Max torque, 1.8kg/7,000rpm. CDI. Electric starter. Lubrication, forced, wet sump. Carburettor, Keihin CV32. Valve timing: Inlet, open 32°BTDC, close 60°ABDC. Exhaust, open 67°BBDC, close 25°ATDC. Five speed gearbox. Primary reduction: 3.285:1. Internal ratios: 2.636, 1.733, 1.300, 1.050 and 0.904:1 (top gear). Clutch, wet multi-plate. Frame: Tubular, single cradle. Front telescopic fork, rear swing arm. Tyres, 2.75 x 18 front, 4.60S x 16 rear. Castor angle 27.5°. Trail, 98mm. Brakes, front drum, 180mm, rear 130mm. Electrical equipment: 12v/10ah battery with 35w headlight. Dimensions: Length, 1,990mm. Width, 710mm. Wheelbase, 1,310mm. Ground clearance, 135mm. Seat height, 755mm. Dry weight, 129kg. Fuel-tank capacity, 9.3 litres.

Kawasaki Z250LTD G2

Engine: Four stroke, single cylinder sohc, 246cc (70 x 64mm). Compression ratio, 8.9:1. Max power, 19hp/8,000rpm. Max torque, 1.8kg/7,000rpm. CDI. Electric starter. Lubrication, forced, wet sump. Carburettor, Keihin CV32. Valve timing: Inlet, open 32°BTDC, close 60°ABDC. Exhaust, open 67°BBDC, close 25°ATDC. Five speed gearbox. Primary reduction: 3.285:1. Internal ratios: 2.636, 1.733, 1.300, 1.050 and 0.904:1 (top gear). Clutch, wet multi-plate. Frame: Tubular, single cradle. Front, telescopic fork, rear swing arm. Tyres, 2.75 x 18 front, 4.60 x 16 rear. Castor angle, 29°. Trail 106mm. Brakes, front drum, 180mm, rear

130mm. Electrical equipment: 12v/10ah battery with 35w headlight. Dimensions: Length, 2,005mm. Width, 810mm. Wheelbase, 1,335mm. Ground clearance, 150mm. Seat height, 725mm. Dry weight, 129kg. Fuel-tank capacity, 8 litres.

Kawasaki Z200-A4

Engine: Four stroke, single cylinder sohc, 198cc (66 x 58mm). Compression ratio, 9.0:1. Max power, 18hp/8,000rpm. Max torque, 1.68kg/7,000rpm. CDI. Electric starter. Lubrication, forced, wet sump. Carburettor, Keihin PW26. Valve timing: Inlet, open 32°BTDC, close 60°ABDC. Exhaust, open 67°BBDC, close 25°ATDC. Five speed gearbox. Primary reduction: 3.285:1. Internal ratios: 2.636, 1.733, 1.300, 1.050 and 0.904:1 (top gear). Clutch, wet multi-plate. Frame: Tubular, single cradle. Front telescopic fork, rear swing arm. Tyres, 2.75 x 18 front, 3.25 x 17 rear. Castor angle, 26°. Trail, 88mm. Brakes, front single disc, rear drum. Electrical equipment: 12v/10ah battery with 35w headlight. Dimensions: Length, 1,980mm. Width, 700mm. Wheelbase, 1,280mm. Ground clearance, 150mm. Seat height, 770mm. Dry weight, 126kg. Fuel-tank capacity, 9.3 litres.

Kawasaki KE175-D3

Engine: Two stroke, single cylinder, reed valve, 174cc (62.5 x 57mm). Compression ratio, 6.5:1. Max power, 16hp/6,500rpm. Max torque, 1.9kg/5,500rpm. CDI. Kick starter. Lubrication, by injection. Carburettor, Mikuni VM26SS. Valve timing: Inlet, open 121°BTDC, close 121°ATDC. Scavenging, open 54°BBDC, close 54°ABDC. Exhaust, open 80°BBDC, close 80°ABDC. Five speed gearbox. Primary reduction: 3.125:1. Internal ratios: 2.666, 1.647, 1.222, 0.950 and 0.800:1 (top gear). Clutch, wet multi-plate. Frame: Tubular, single cradle. Front telescopic fork, rear swing arm. Tyres, 2.75 x 21 front, 3.50 x 18 rear. Castor angle, 29°. Trail,

116mm. Brakes, drum, 120mm front and rear. Electrical equipment: 6v/6ah battery with 35w headlight. Dimensions: Length, 2,130mm. Width, 980mm. Wheelbase, 1,360mm. Ground clearance, 245mm. Seat height, 845mm. Dry weight, 102kg. Fuel-tank capacity, 9.6 litres.

Kawasaki KH125-A4

Engine: Two stroke, single cylinder, rotary disc valve, 124cc (56 x 50.6mm). Compression ratio, 7.0:1. Max power, 13.5hp/7,000rpm. Max torque, 1.43kg/5,500rpm. CDI, Kick starter. Lubrication, by injection. Carburettor, Mikuni VM24SS. Valve timing: Inlet, open 115°BTDC, close 55°ATDC. Scavenging, open 56.5°BBDC, close, 56.5°ABDC. Exhaust, open 84.5°BBDC, close 84.5°ABDC. Six speed gearbox. Primary reduction: 3.136:1. Internal ratios, 2.600, 1.692, 1.250, 1.045, 0.894 and 0.800:1 (top gear). Clutch, wet multi-plate. Frame: Tubular, single cradle. Front telescopic fork, rear swing arm. Tyres, 2.75 x 18 front, 3.00 x 18 rear. Castor angle, 24.5°. Trail, 74mm. Brakes, front single mechanical disc, 19mm, rear, 110mm. Electrical equipment: 6v/6ah battery with 35w headlight. Dimensions: Length, 1,900mm. Width, 650mm. Wheelbase, 1,235mm. Ground clearance, 170mm. Seat height, 785mm. Dry weight, 95kg. Fuel-tank capacity, 11.5 litres.

Kawasaki KE125-A8

Engine: Two stroke, single cylinder, rotary disc valve, 124cc (56 x 50.6mm). Compression ratio, 7.0:1. Max power, 12.5hp/6,500rpm. Max torque, 1.42kg/6,000rpm. CDI. Kick starter. Lubrication, by injection. Carburettor, Mikuni VM24SS. Valve timing: Inlet, open 115°BTDC, close 55°ATDC. Scavenging, open 56°BBDC, close 56°ABDC. Exhaust, open 80°BBDC, close 80°ABDC. Six speed gearbox. Primary reduction: 3.136:1. Internal ratios: 2.600, 1.692, 1.250, 1.045, 0.894 and 0.800:1 (top gear).

Clutch, wet multi-plate. Frame: Tubular, single cradle. Front telescopic fork, rear swing arm. Tyres, 2.75 x 21 front, 3.50 x 18 rear. Castor angle, 30°. Trail, 127mm. Brakes, front drum, 120mm, rear, 130mm. Electrical equipment: 6v/6ah battery with 35w headlight. Dimensions: Length, 2,100mm. Width, 845mm. Wheelbase, 1,350mm. Ground clearance, 275mm. Seat height, 845mm. Dry weight, 99kg. Fuel-tank capacity, 9.6 litres.

Kawasaki KE100-A10

Engine: Two stroke, single cylinder, rotary disc valve, 99cc (49.5 x 51.8mm). Compression ratio, 7.0:1. Max power, 11hp/7,500rpm. Max torque, 1.1kg/7,000rpm. CDI. Kick starter. Lubrication, by injection. Carburettor, Mikuni VM19SC. Valve timing: Inlet, open 120°BTDC, close 55°ATDC. Scavenging, open 58°35'BBDC, close 58°35'ABDC. Exhaust, open 84°16'BBDC, close 84°16'ABDC. Five speed gearbox. Primary reduction: 3.523:1. Internal ratios: 2.916, 1.764, 1.300, 1.090 and 0.928:1 (top gear). Clutch, wet multi-plate. Frame: Tubular, double cradle. Front telescopic fork, rear swing arm. Tyres, 2.75 x 19 front, 3.00 x 18 rear. Castor angle, 29.5°. Trail, 120mm. Brakes, drum, 110mm, front and rear, Electrical equipment: 6v/4ah battery with 25w headlight. Dimensions: Length, 1,980mm. Width, 860mm. Wheelbase, 1,260mm. Ground clearance, 240mm. Seat height, 805mm. Dry weight, 92kg. Fuel-tank capacity, 8 litres.

Kawasaki KC100-C2

Engine: Two stroke, single cylinder, rotary disc valve, 99cc (49.5 x 51.8mm). Compression ratio, 7.0:1. Max power, 10.5hp/7,500rpm. Max torque, 1.0kg/7,000rpm. CDI. Kick starter. Lubrication, by injection. Carburettor, Mikuni VM19SC. Valve timing: Inlet, open 120°BTDC, close 55°ATDC. Scavenging, open 58°35'BBDC, close 58°35'ABDC.

Exhaust, open 84°16'BBDC, close 84°16'ABDC. Five speed gearbox. Primary reduction: 3.523:1. Internal ratios: 2.916, 1.764, 1.300, 1.090 and 0.928:1 (top gear). Clutch, wet multi-plate. Frame: Tubular, double cradle. Front telescopic fork, rear swing arm. Tyres, 2.50 x 18 front and rear. Castor angle, 26°. Trail, 80mm. Brakes, drum, 110mm, front and rear. Electrical equipment: 6v/4ah battery with 25w headlight. Dimensions: Length, 1,810mm. Width, 740mm. Wheelbase, 1,150mm. Ground clearance, 150mm. Seat height, 765mm. Dry weight, 82kg. Fuel-tank capacity, 8.6 litres.

Kawasaki KM100-A7

Engine: Two stroke, single cylinder, rotary disc valve, 99cc (49.5 x 51.8mm). Compression ratio, 7.2:1. Max power, 8.5hp/6,500rpm. Max torque, 0.96kg/5,500rpm. CDI. Kick starter. Lubrication, by injection. Carburettor, Mikuni VM19SC. Valve timing: Inlet, open 120°BTDC, close 55°ATDC. Scavenging, open 55°BBDC, close 55°ABDC. Exhaust, open 78°BBDC, close 78°ABDC. Five speed gearbox. Primary reduction: 3.523:1. Internal ratios: 2.916, 1.764, 1.300, 1.090 and 0.958:1 (top gear). Clutch, wet multi-plate. Frame: Tubular, single cradle. Front telescopic fork, rear swing arm. Tyres, 2.50 x 16 front, 3.00 x 14 rear. Castor angle, 26.5°. Trail, 70mm. Brakes, drum, 110mm, front and rear. Electrical equipment: 6v/4ah battery with 25w headlight. Dimensions: Length, 1,720mm. Width, 765mm. Wheelbase, 1,110mm. Ground clearance, 185mm. Seat height, 720mm. Dry weight, 77kg. Fuel-tank capacity, 6.5 litres.

Kawasaki AR80-A1

Engine: Two stroke, single cylinder, piston reed valve, 78cc (49 x 41.6mm). Compression ratio, 7.8:1. Max power, 10hp/8,000rpm. Max torque, 0.89kg/7,500rpm. CDI. Kick starter. Lubrication, by injection. Carburettor, Mikuni VM18SS. Valve timing: Inlet, full open. Scavenging, open 56°BBDC, close 56°ABDC, Exhaust, open 86°BBDC, close 86°ABDC. Six speed gearbox. Primary reduction: 3.619:1. Internal ratios: 3.307, 2.111, 1.545, 1.240, 1.074 and 0.965:1. Clutch, wet multi-plate. Frame: Tubular, semi double cradle. Front telescopic fork, rear swing arm. Tyres, 2.50 x 18 front, 2.75 x 18 rear. Castor angle, 27°15'. Trail, 83mm. Brakes, front single disc, 182mm, rear drum, 110mm. Electrical equipment: 6v/6ah battery with 25w headlight. Dimensions: Length, 1,855mm. Width, 630mm. Wheelbase, 1,205mm. Ground clearance, 175mm. Seat height, 790mm. Dry weight, 75kg. Fuel-tank capacity, 9.6 litres.

Kawasaki AR50-A1

The AR50-A (for 16-year-olds) looks very like a full-size Enduro, with a leading-axle front fork and Uni-Trak rear suspension offering 120mm of travel. The engine has reed-valve induction, the tank is impressively humpy, and, all told, the bike has plenty of 'presence' despite its small size.

Engine: Two stroke, single cylinder, piston reed valve, 49cc (39 x 41.6mm). Compression ratio, 7:1. Max power, 2.9hp/4,500rpm. Max torque, 0.46kg/4,000rpm. CDI. Kick starter. Lubrication, by injection. Carburettor, Mikuni VM14SS. Valve timing: Inlet, full open. Scavenging, open 50°BBDC, close 50°ABDC. Exhaust, open 68°BBDC, close 68°ABDC. Five speed gearbox. Primary reduction: 3.619:1. Internal ratios: 3.307, 2.111, 1.545, 1.240 and 1.074:1. Clutch, wet multi-plate. Frame: Tubular, semi double cradle. Front tele-scopic fork, rear pivoted fork with spring units. Tyres, 2.50 x 18 front, 2.75 x 18 rear. Castor angle, 27°30'. Trail, 85mm. Brakes, front single disc, 182mm, rear drum, 110mm. Electrical equipment: 6v/6ah battery with 25w headlight. Dimensions: Length, 1,855mm. Width, 630mm. Wheelbase, 1,195mm. Ground clearance, 175mm. Seat height, 790mm. Dry weight, 75kg. Fuel-tank capacity, 9.6 litres.

SUZUKI

The Katana models are the leaders of the latest Suzuki range. They are available as GSX1100S, GS650G and GS550M. It is the styling of the Katanas — the integrated tank/seat treatment and the short, crammed look to the engine/gearbox unit — rather than the specification, that makes them stand out from the herd. Power output varies from 100bhp for the 1100 to 54 for the 550, with transistorized ignition, and five- and six-speed transmission. The styling is a very considered attempt to achieve a fresh appearance for the contemporary motorcycle. While it does not take the easy option of complete enclosure — wisely, because the experiences of manufacturers who have gone this way have not been very happy — the halfway house approach of Suzuki is a good compromise.

Suzuki GSX1100S

Engine: Four stroke, four cylinder dohc, 16 valve, 1075cc (72 x 66mm). Compression ratio, 9.5:1. Max power, 99.6hp/8,700rpm. Electric starter. Five speed gearbox. 12v battery. Tyres, 3.50 x 19 front, 4.50 x 17 rear. Dry weight, 241kg.

Suzuki GSX1000

Big Suzukis have a high reputation not merely for speed, which is around 128 mph, maximum, but for all-round handling somewhat above average in this generally overweight cc division. Though perhaps not much more than 20-30 lb lighter than other makers' thousands, the Suzukis, with a good frame, are reckoned to be outstanding raw material for go-quicker 'specials'.

Engine: Four stroke, four cylinder dohc, 997cc (70 x 64.8mm). Compression ratio, 9.2:1. Max power, 89.5hp/8,500rpm. Electric starter. Five speed gearbox. 12v battery. Tyres, 3.50 x 19 front, 4.50 x 17 rear. Dry weight, 255kg.

Suzuki GSX750

Engine: Four stroke, four cylinder dohc, 747cc (67 x 53mm). Compression ratio, 9.4:1. Max power, 80hp/9,200rpm. Electric starter. Five speed gearbox. 12v battery. Tyres, 3.25 x 19 front, 4.00 x 18 rear. Dry weight, 233kg.

Suzuki GSX400F

Four-stroke, four valves per cylinder, four hundred cc . . . the GSX400F has plenty of performance — a little more edge than the equivalent Kawasaki. It was introduced in Europe in 1981.

Engine: Four stroke, air-cooled, four cylinder dohc, 16 valve, 399cc (53 x 45.2mm). Compression ratio, 10:1. Max power, 40hp/9,500rpm. Electric starter. Six speed gearbox. 12v battery. Tyres, 3.25 x 19 front, 3.75 x 18 rear. Dry weight, 181kg.

Suzuki GSX400T

Another 400, the GSX400T, has two cylinders (and eight valves) but produces much the same power as the four . . . something to do, perhaps, with the 'Twin Swirl Combustion Chambers' of which Suzuki are so proud. Where the four has three disc brakes, the twin has a drum unit at the rear and a single disc at the front. The differences in equipment reflect the contrasting images Suzuki are keen to promote for the bikes; performance, however, is on a par.

Engine: Four stroke, two cylinder dohc, 399cc (67 x 56.6mm). Compression ratio, 10:1. Max power, 40.8hp/9,000rpm. Electric starter. Six speed gearbox. 12v battery. Tyres, 3.60 x 19 front, 4.60 x 16 rear. Dry weight, 174kg.

Suzuki GSX400

Engine: Four stroke, two cylinder dohc, 399cc (67 x 56.6mm). Compression ratio, 10:1. Max power, 41.56hp/9,000rpm. Electric starter. Six speed gearbox. 12v battery. Tyres, 3 x 18 front, 3.50 x 18 rear. Dry weight, 175kg.

Suzuki GSX250

Engine: Four stroke, two cylinder dohc, 8 valve, 249cc (60 x 44.2mm). Compression ratio, 10.5:1. Max power, 27hp/10,000rpm. Electric

starter. Six speed gearbox. 12v battery. Tyres, 3 x 18 front, 3.75 x 18 rear. Dry weight, 159kg.

Suzuki GS1000ET

Engine: Four stroke, four cylinder dohc, 997cc (70 x 64.8mm). Compression ratio, 9.2:1. Max power, 89hp/8,500rpm. Electric starter. Five speed gearbox. 12v battery. Tyres, 3.50 x 19 front, 4.50 x 17 rear. Dry weight, 234kg.

Suzuki GS650G

Engine: Four stroke, four cylinder dohc, 674cc (62 x 55.8mm). Compression ratio, 9.5:1. Max power, 66hp/9,000rpm. Electric starter. Five speed gearbox. 12v battery. Tyres, 3.25 x 19 front, 3.75 x 18 rear. Dry weight, 218kg.

Suzuki GS650GT

Engine: Four stroke, four cylinder dohc, 647cc (62 x 55.8mm). Compression ratio, 9.5:1. Max power, 60hp/9,000rpm. Electric starter. Five speed gearbox. 12v battery. Tyres, 3.25 x 19 front, 4.25 x 17 rear. Dry weight, 215kg.

Suzuki GS550M

Engine: Four stroke, four cylinder dohc, 549cc (56 x 55.8mm). Compression ratio, 8.6:1. Max power, 53.7hp/9,000rpm. Electric starter. Six speed gearbox. 12v battery. Tyres, 3.25 x 19 front and 3.75 x 18 rear. Dry weight, 205kg.

Suzuki GS550

Engine: Four stroke, four cylinder dohc, 549cc (56 x 55.8mm). Compression ratio, 8.6:1. Max power, 53.7hp/9,000rpm. Electric starter. Six speed gearbox. 12v battery. Tyres, 3.25 x 19 front, 3.75 x 18 rear. Dry weight, 205kg.

Suzuki GS450

Engine: Four stroke, two cylinder dohc, 448cc (71 x 56.6mm). Compression ratio, 10:1. Max power, 42.3hp/9,000rpm. Electric starter. Six speed gearbox. 12v battery. Tyres, 3 x 18 front, 3.50 x 18 rear. Dry weight, 175kg.

Suzuki GS250TT

Engine: Four stroke, two cylinder dohc, 249cc (60 x 44.2mm). Compression ratio, 10.5:1. Max power, 27hp/10,000rpm. Electric starter. Six speed gearbox. 12v battery. Tyres, 3 x 18 front, 3.50 x 17 rear. Dry weight, 160kg.

Suzuki GN400

Engine: Four stroke, single cylinder, 396cc (88 x 65.2mm). Compression ratio, 9.2:1. Max power, 27.6hp/7,200rpm. Kick

Suzuki TS185

Engine: Two stroke, single cylinder, 185cc (64 x 57mm). Compression ratio, 6.2:1. Max power, 16.5hp/5,500rpm. Kick starter. Five speed gearbox. 6v battery. Tyres, 2.75 x 21 front, 4.10 x 18 rear. Dry weight, 102kg.

Suzuki TS125

Engine: Two stroke, single cylinder, 123cc (56 x 50mm). Compression ratio, 6.6:1. Max power, 13.5hp/8,000rpm. Kick starter. Six speed gearbox. 6v battery. Tyres, 2.75 x 21 front, 4.10 x 18 rear. Dry weight, 97kg.

Suzuki TS100

Engine: Two stroke, single cylinder, 98cc (50 x 50mm). Compression ratio, 6.4:1. Max power, 10.5hp/8,000rpm. Kick starter. Five speed gearbox. 6v battery. Tyres, 2.75 x 19 front, 2.75 x 18 rear. Dry weight, 91kg.

starter. Five speed gearbox. Tyres, 3.60 x 18 front, 4.60 x 16 rear. Dry weight, 143kg.

Suzuki SP400

Among the smaller Suzukis there is a decent split between two- and four-stroke, with the two-strokes predominating at the 100cc-and-below extreme. Dual-purpose — for off-road and ordinary use — machinery is available in some variety as SP and DR 400, and in plenty of smaller sizes as TS250, 185, 125 and 100.

Engine: Four stroke, single cylinder sohc, 396cc (88 x 65.2mm). Compression ratio, 9.2:1. Max power, 25hp/7,500rpm. Kick starter. Five speed gearbox. 6v battery. Tyres, 3.00 x 21 front, 4.00 x 18 rear. Dry weight, 127kg.

Suzuki DR400

Engine: Four stroke, single cylinder sohc, 396cc (88 x 65.2mm). Compression ratio, 9.3:1. Max power, 27hp/7,500rpm. Kick starter. Five speed gearbox. 6v battery. Tyres, 3.00 x 21 front, 4.60 x 18 rear. Dry weight, 130.5kg.

Suzuki TS250

Engine: Two stroke, single cylinder, 246cc (70 x 64mm). Compression ratio, 5.9:1. Max power, 23hp/6,000rpm. Kick starter. Five speed gearbox. 6v battery. Tyres, 3.00 x 21 front, 4.60 x 18 rear. Dry weight, 121kg.

Suzuki TS50

Engine: Two stroke, single cylinder, 49cc (41 x 37.8mm). Compression ratio, 6.7:1. Max power, 2.92hp/6,000rpm. Kick starter. Five speed gearbox. 6v battery. Tyres, 2.75 x 21 front, 3.00 x 18 rear. Dry weight, 82kg.

Suzuki TS125 Farmbike

Engine: Two stroke, single cylinder, 123cc (56 x 50mm). Compression ratio, 6.6:1. Max power, 12.5hp/8,000rpm. Kick starter. Six speed gearbox. 6v battery. Tyres, 2.75 x 21 front, 3.50 x 18 rear. Dry weight, 97.5kg.

Suzuki GT250X7

Engine: Two stroke, two cylinder, 247cc (54 x 54mm). Compression ratio, 6.7:1. Max power, 29hp/8,000rpm. Kick starter. Six speed gearbox. Tyres, 3 x 18 front, 3.50 x 18 rear. Dry weight, 128kg.

Suzuki GT200X5

Engine: Two stroke, two cylinder, 196cc (50 x 50mm). Compression ratio, 7:1. Max power, 20hp/8,000rpm. Electric and kick starter. Five speed gearbox. Tyres, 2.75 x 18 front, 3 x 18 rear. Dry weight, 123kg.

Suzuki GT125

Engine: Two stroke, two cylinder, 124cc (43 x 43mm). Compression ratio, 6.8:1. Max power, 16hp/9,500rpm. Kick starter. Five speed gearbox. 12v battery. Tyres, 2.75 x 18 front, 3.00 x 18 rear. Dry weight, 108kg.

Suzuki GP125

Engine: Two stroke, single cylinder disc valve, 123cc (56 x 50mm). Compression ratio, 6.7:1. Max power, 15hp/8,500rpm. Kick starter. Five speed gearbox. 6v battery. Tyres, 2.75 x 18 front, 3 x 18 rear. Dry weight, 92kg.

Suzuki SB200

Engine: Two stroke, two cylinder, 196cc (50 x 50mm). Compression ratio, 6.5:1. Max power, 18.3hp/7,500rpm. Kick starter. Four speed gearbox. 12v battery. Tyres, 2.75 x 18 front, 3 x 18 rear. Dry weight, 116kg.

Suzuki GP100 Sport

Engine: Two stroke, single cylinder disc valve, 98cc (50 x 50mm). Compression ratio, 6.8:1. Max power, 12hp/8,500rpm. Kick starter. Five speed gearbox. 6v battery. Tyres, 2.50 x 18 front, 2.75 x 18 rear. Dry weight, 89kg.

Suzuki FR80

Engine: Two stroke, single cylinder, 79cc (49 x 42mm). Compression ratio, 6.7:1. Max power, 6.8hp/6,500rpm. Kick starter. Three speed gearbox. Automatic clutch. 6v battery. Tyres, 2.25 x 17 front and rear. Dry weight 73kg.

Suzuki OR50

Engine: Two stroke, single cylinder, 49cc (41 x 37.4mm). Compression ratio, 6.7:1. Max power, 2.4hp/5,000rpm. Kick starter. Five speed gearbox. 6v battery. Tyres, 2.50 x 17 front, 3.00 x 14 rear. Dry weight, 67kg.

Suzuki ZR50X1

Engine: two stroke, single cylinder, 49cc (41 x 37.8mm). Compression ratio, 6.7:1. Max power, 2.92hp/6,000rpm. Kick starter. Five speed gearbox. 6v battery. Tyres, 2.50 x 18 front and rear. Dry weight, 89kg.

Suzuki FS50

Engine: Two stroke, single cylinder, 49cc (41 x 37.8mm). Compression ratio, 6.1:1. Max power, 2.52hp/6,000rpm. Kick starter. Two speed automatic gearbox. 6v battery. Tyres, 2.75 x 10 front and rear. Dry weight, 53.5kg.

Suzuki FZ50

Engine: Two stroke, single cylinder, 49cc (41 x 37.8mm). Compression ratio, 6.1:1. Max power, 2.5hp/6,000rpm. Kick starter. Two speed gearbox. 6v battery. Tyres, 3.00 x 12 front and rear. Dry weight, 59kg.

Suzuki ZR50L

Engine: Two stroke, single cylinder, 49cc (41 x 37.8mm). Compression ratio, 6.8:1. Max power, 2.8hp/6,500rpm. Kick starter. Five speed gearbox. Tyres, 2.50 x 19 front and 3.50 x 16 rear. Dry weight, 92kg.

Suzuki FR50

Engine: Two stroke, single cylinder, 49cc (41 x 37.8mm). Compression ratio, 7:1. Max power, 3hp/5,500rpm. Kick starter. Three speed gearbox with automatic clutch. 6v battery. Tyres, 2.25 x 17 front and rear. Dry weight, 70kg.

Suzuki GP100U

Engine: Two stroke, single cylinder disc valve, 98cc (50 x 50mm). Compression ratio, 6.8:1. 12hp/8,500rpm. Kick starter. Five speed gearbox. 6v battery. Tyres, 2.50 x 18 front, 2.75 x 18 rear. Dry weight, 86kg.

YAMAHA

Yamaha XS1100

XS1100 — A while back, cynics used to spell out XS as Excess, but that was before Kawasaki's Z1300 lumbered on to the scene. A big seller in the US where its price is much below that of the biggest BMW, the 1100 makes a fine tourer. It is not, however, over handy in tight-cornering situations; but 600 lb long-wheelbase motor cycles have seldom excelled in this area.

Engine: Four stroke, four cylinder dohc, 1,101cc (71.5 x 68.6mm). Compression ratio, 9.2:1. Max power 95hp/8,500rpm. Max torque, 9.2kg/6,000rpm. CDI. Electric/kick starter. Lubrication, pressure fed, wet sump. Carburettor, BS3411 x 4. Five speed gearbox. Tyres, 3.50 x 19 front, 4.50 x 17 rear. Brakes, front dual discs, rear single disc. Dimensions: Length, 2,260mm. Width, 775mm. Wheelbase, 1,545mm. Ground clearance, 150mm. Seat height, 810mm. Dry weight, 256kg. Fuel-tank capacity, 24 litres.

Yamaha XV750SE

The new vee-twins move sharply away from the hitherto 'traditional' Japanese practice of high revs from three, preferably four, cylinders. All credit to Yamaha as the company to recognize the market that exists for vees — small, perhaps, but worthwhile.

Engine: Four stroke, V-twin cylinder sohc, 748cc (83 x 69.2mm). Compression ratio, 8.7:1. Max power, 60.9hp/7,000rpm. Max torque, 6.6kg/6,000rpm. CDI Electric starter. Lubrication, wet sump. Five speed gearbox. Tyres, 3.50 x 19 front, 130/90 x 16 rear. Brakes, disc front, drum rear. Dimensions: Length, 2,230mm. Width, 840mm. Wheelbase, 1,520mm. Ground clearance, 145mm. Seat height, 745mm. Dry weight 212kg. Fuel-tank capacity 12 litres.

Yamaha TR1

Engine: Four stroke, V-twin cylinder sohc, 981cc (95 x 69.2mm). Compression ratio, 8.3:1. Max

power, 70hp/6,500rpm. Max torque, 8.28kg/5,500rpm. CDI. Electric starter. Lubrication, wet sump. Five speed gearbox. Tyres, 3.25 x 19 front, 120 x 18 rear. Brakes, disc front, drum rear. Dimensions: Length, 2,265mm. Width, 730mm. Wheelbase, 1,540mm. Ground clearance, 140mm. Seat height, 770mm. Dry weight, 220kg. Fuel-tank capacity, 19 litres.

Yamaha XS850

Still notable, after several years, as the only three-cylinder four-stroke in production, the XS750 has good performance and a certain style that is lacking in its bigger stablemate, the XS850, another 'shaftie'.

Engine: Four stroke, three cylinder dohc, 826cc (71.5 x 68.6mm). Compression ratio, 9.2:1. Max power, 79hp/8,500rpm. Max torque, 7.1kg/7,500rpm. CDI. Electric starter. Lubrication, pressure fed, wet sump, with oil cooler. Carburettor, LD120. Primary transmission, chain/final, shaft. Five speed gearbox. Tyres, 3.25 x 19 front, 4.00 x 18 rear. Brakes, hydraulic disc front and rear. Dimensions: Length, 2,155mm. Width, 675mm. Wheelbase, 1,465mm. Ground clearance, 130mm. Seat height, 815mm. Dry weight, 236kg. Fuel-tank capacity, 24 litres.

Yamaha XS750SE

Some makers use 'LTD' to distinguish their US-orientated models: Yamaha, however, go for 'SE'. Either way, the main differences from the remainder of the range, now presumably acknowledged as 'European' in styling and purpose, are the high, wide handlebars and the lower, stepped dual-seat.

Engine: Four stroke, three cylinder dohc, 747cc (68 x 68.6mm). Compression ratio, 9.2:1. Max power, 68hp/8,000rpm. CDI. Electric starter. Lubrication, wet sump. Five speed gearbox. Tyres, 3.25 x

19 front, 4.00 x 18 rear. Brakes, dual disc front, single disc rear. Dimensions: Length, 2,155mm. Wheelbase, 1,485mm. Ground clearance, 155mm. Seat height, 820mm. Dry weight, 236kg. Fuel-tank capacity, 17 litres.

Yamaha XS650SE

Engine: Four stroke, two cylinder sohc, 653cc (75 x 74mm). Compression ratio, 9.2:1. Max power, 50.1hp/7,000rpm. CDI. Electric/kick starter. Lubrication, wet sump. Five speed gearbox. Tyres, 3.50 x 19 front, 130/90 x 16 rear. Brakes, disc front and rear. Dimensions: Length, 2,130mm. Wheelbase, 1,435mm. Ground clearance, 135mm. Dry weight, 212kg. Fuel-tank capacity, 11.5 litres.

Yamaha XJ650

Engine: Four stroke, four cylinder dohc, 653cc (63 x 52.4mm). Compression ratio, 9.2:1. Max power, 73hp/9,000rpm. Max torque, 6.0kg/7,500rpm. CDI. Electric starter. Lubrication, wet sump. Carburettor, HSC28. Shaft drive. Front telescopic fork, rear pivoted fork. Tyres, 3.25 x 19 front, 120/90 x 18 rear. Brakes, double disc front, drum rear. Dimensions: Length, 2,170mm. Width, 730mm. Wheelbase, 1,435mm. Ground clearance, 140mm. Seat height, 780mm. Dry weight, 206kg. Fuel-tank capacity, 19.5 litres.

Yamaha SR500

Engine: Four stroke, single cylinder sohc, 499cc (87 x 84mm). Compression ratio, 9:1. Max power 33hp/6,500rpm. Max torque, 3.9kg/5,500rpm. CDI. Kick starter. Lubrication, dry sump. Carburettor, VM34SS. Five speed gearbox. Front telescopic fork, rear pivoted fork. Tyres, 3.50S x 19 4PR front, 4.00S x 18 4PR rear. Brakes,

disc front, drum rear. Dimensions: Length, 2,105mm. Width, 930mm. Wheelbase, 1,400mm. Seat height, 810mm. Dry weight, 158kg. Fuel-tank capacity, 12 litres.

Yamaha XS400SE

Engine: Four stroke, two cylinder sohc, 391cc (69 x 52.4mm). Max power, 37hp/9,000rpm. CDI. Electric/kick starter. Lubrication, wet sump. Six speed gearbox. Brakes, disc front and rear. Dimensions: Length, 2,065mm. Wheelbase, 1,380mm. Ground clearance, 135mm. Seat height, 770mm. Dry weight, 169kg. Fuel-tank capacity, 14 litres.

Yamaha RD250LC

In the UK the Yamaha-sponsored Pro-Am race series caters for 24-year-old (and younger) international riders in two categories — the Pros, riders of proven ability and the Ams, apprentice racers graduating from club competition. Riders in both classes are on RD/LC models (of 350cc), which is possibly all that has to be said about these liquid-cooled 100 mph-plus lightweights.

Engine: Two stroke, two cylinder, liquid cooled, 247cc (54 x 54mm). Compression ratio, 7:1. Max power, 35.5hp/8,500rpm. CDI. Lubrication, Autolube. Carburettor, Mikuni VM26SS. Six speed gearbox. Clutch, wet multi-plate. Tyres, 3 x 18 front, 3.50 x 18 rear.

Brakes, disc front, drum rear. Dimensions: Length, 2,055mm. Wheelbase, 1,360mm. Ground clearance, 170mm. Seat height, 785mm. Dry weight, 139kg. Fuel-tank capacity, 17 litres.

Yamaha XS250

Engine: Four stroke, two cylinder sohc, 248cc (55 x 52.4mm). Compression ratio, 9.3:1. Max power, 27hp/9,500rpm. Electric/kick starter. Lubrication, pressure-fed, wet sump. Carburettor, BS32 x 2. Brakes, hydraulic disc front, drum rear. Six speed gearbox. Clutch, wet multi-plate. Tyres, 3 x 18 front, 3.75 x 18 rear. Dimensions: Length, 2,080mm. Wheelbase, 1,380mm. Ground clearance, 150mm. Seat height, 780mm. Dry weight, 168kg. Fuel-tank capacity, 17 litres.

Yamaha XS250SE

Engine: Four stroke, twin cylinder sohc, 248cc (55 x 52.4mm). Max power, 26hp/8,500rpm. Electric/kick starter. Lubrication, wet sump. Six speed gearbox. Tyres, 3 x 18 front, 120/90 x 16 rear. Brakes, disc front, drum rear. Dimensions: Length, 2,065mm. Seat height, 770mm. Wheelbase, 1,375mm. Ground clearance, 135mm. Dry weight, 169kg. Fuel-tank capacity, 14 litres.

Yamaha SR250SE

Engine: Four stroke, single cylinder sohc, 239.6cc (73.5 x 56.5mm). Max power, 17hp/7,500rpm. Electric starter. Five speed gearbox. Tyres, 3 x 19 front, 120/90 x 16 rear. Dimensions: Length, 2,025mm. Wheelbase, 1,335mm. Ground clearance, 145mm. Seat height, 740mm. Fuel-tank capacity, 11 litres.

Yamaha RS125

Engine: Two stroke, single cylinder, 123cc (56 x 50mm). Compression ratio. 6.9:1. Max power, 14hp/7,500rpm. Lubrication, Autolube. Primary kick starter. Carburettor, VM24SH. Five speed gearbox. Clutch, wet multi-plate. Tyres, 2.75 x 18 front, 3 x 18 rear. Brakes, front hydraulic disc, drum rear. Dimensions: Length, 1,945mm. Wheelbase, 1,240mm. Seat height, 775mm. Dry weight, 96kg. Fuel-tank capacity, 10 litres.

Yamaha RS100

Engine: Two stroke, single cylinder, torque induction, 97cc (52 x 45.6mm). Compression ratio, 7:1. Max power, 11hp/8,500rpm. Lubrication, Autolube. Primary kick starter. Carburettor, VM20SH. Five speed gearbox. Clutch, wet multi-plate. Tyres, 2.75 x 18 front, 3 x 18 rear. Brakes, drum front and rear. Dimensions: Length, 1,945mm. Wheelbase, 1,240mm. Seat height, 775mm. Dry weight, 91kg. Fuel-tank capacity, 10 litres.

Yamaha DT175MX

Engine: Two stroke, single cylinder, 171cc (66 x 50mm). Compression ratio, 6.8:1. Max power, 16.3hp/7,000rpm. Primary kick starter. Lubrication, Autolube. Carburettor, VM24. Six speed gearbox. Clutch, wet multi-plate. Tyres, 2.75 x 21 front, 3.50 x 18 rear. Brakes, drum front and rear. Dimensions: Length, 2,080mm. Wheelbase, 1,350mm. Ground clearance, 265mm. Seat height, 855mm. Dry weight, 99kg. Fuel-tank capacity, 7 litres.

Yamaha DT125MX

Engine: Two stroke, single cylinder torque induction, 123cc (56 x 50mm). Compression ratio, 7.2:1. Max power, 14hp/6,500rpm. Kick starter. Lubrication, Autolube. Carburettor, VM22. Six speed gearbox. Clutch, wet multi-plate. Tyres, 2.75 x 21 front, 3.50 x 18 rear. Brakes, drum front and rear. Dimensions: Length, 2,080mm. Wheelbase, 1,350mm. Ground clearance, 265mm. Seat height, 855mm. Dry weight, 96kg. Fuel-tank capacity, 7 litres.

Yamaha DT100

Engine: Two stroke, single cylinder torque induction, 96cc (52 x 45.6mm). Compression ratio, 7.2:1. Max power, 10hp/7,500rpm. Kick starter. Lubrication, Autolube. Carburettor, VN22SS. Five speed gearbox. Clutch, wet multi-plate. Tyres, 2.75 x 19 front, 3 x 18 rear. Brakes, drum front and rear. Dimensions: Length, 1,960mm. Wheelbase, 1,305mm. Ground clearance, 240mm. Seat height, 795mm. Dry weight, 94kg. Fuel-tank capacity, 7 litres.

Yamaha DT50M

Engine: Two stroke, single cylinder torque induction, 49cc (40 x 39.7mm). Compression ratio, 6.8:1. Kick starter. Lubrication, Autolube. Carburettor, VM16SH. Five speed gearbox. Clutch, wet multi-plate. Tyres, 2.50 x 19 front, 3 x 17 rear. Brakes, drum front and rear. Dimensions: Length, 1,860mm. Wheelbase, 1,210mm. Seat height, 780mm. Dry weight, 72kg. Fuel-tank capacity, 6 litres.

Yamaha XT250

Engine: Four stroke, single cylinder sohc, 249cc (75 x 56.5mm). Compression ratio, 9.2:1. Max power, 21hp/8,000rpm. Primary kick starter. Lubrication, wet sump. Five speed gearbox. Clutch, wet multi-plate. Tyres, 3.00 x 21 front, 4.60 x 17 rear. Brakes, drum front and rear. Dimensions:

Length, 2,135mm. Wheelbase, 1,395mm. Ground clearance, 255mm. Seat height, 840mm. Dry weight, 113kg. Fuel-tank capacity, 8 litres.

Yamaha XT500

Yamaha's trail bikes come as two four-strokes, four two-strokes. The XT500 took the world by storm, say Yamaha — which is not quite true. However the latest XTs, 500 and 250, are good examples of lightweight cross-country bikes having the special characteristics of four-stroke torque.

Engine: Four stroke, single cylinder sohc, 499cc (87 x 84mm). Compression ratio, 9.0:1. Max power, 32hp/6,500rpm. Kick starter. Lubrication, dry sump. Carburettor, VM32SS. Five speed gearbox. Clutch, wet multi-plate. Tyres, 3.25 x 21-4PR front, 4.00 x 18-4PR rear. Brakes, drum front and rear. Dimensions: Length, 2,175mm. Wheelbase, 1,420mm. Ground clearance, 220mm. Seat height, 855mm. Dry weight, 140kg. Fuel-tank capacity, 8.5 litres.

Yamaha IT175

Engine: Two stroke, single cylinder, 171cc (66 x 50mm). Compression ratio, 7.9:1. Max power, 24.5hp/8,000rpm. Max torque, 2.2kg/7,500rpm. Lubrication, pre-mix. Carburettor, Mikuni VM34. Six speed gearbox. Front telescopic forks, monocross rear. Tyres, 3.00 x 21 front, 4.10 x 18 rear. Brakes, drum front and rear. Dimensions: Length, 2,120mm. Width, 870mm. Wheelbase, 1,420mm. Ground clearance, 290mm. Seat height, 895mm. Dry weight, 94kg. Fuel-tank capacity, 11 litres.

Yamaha IT125

Engine: Two stroke, single cylinder, 123cc (56 x 50mm). Compression ratio, 8:1. Max power, 19.5hp/9,000rpm. Max torque, 1.59kg/8,500rpm. CDI. Kick starter. Lubrication, pre-mix.

Carburettor, Mikuni VM30. Six speed gearbox. Front telescopic forks, monocross rear. Tyres, 3 x 31 front, 4.10 x 18 rear. Brakes, drum front and rear. Dimensions: Length, 2,115mm. Width, 860mm. Wheelbase, 1,375mm. Ground clearance, 310mm. Seat height, 860mm. Dry weight, 92kg. Fuel-tank capacity, 8.5 litres.

Yamaha RD50M

Engine: Two stroke, single cylinder, 49cc (40 x 39.7mm). Compression ratio, 6.8:1. Kick starter. Lubrication, Autolube. Carburettor, VM16SH. Five speed gearbox. Clutch, wet multi-plate. Tyres, 2.50 x 18 front, 2.75 x 18 rear. Brakes, front hydraulic disc, drum rear. Dimensions: Length, 1,870mm. Wheelbase, 1,200mm. Seat height, 760mm. Ground clearance, 180mm. Dry weight, 79kg. Fuel-tank capacity, 8 litres.

Yamaha TY50M

Engine: Two stroke, single cylinder, 49cc (40 x 39.7mm). Compression ratio, 6.8:1. Kick starter. Lubrication, Autolube. Carburettor, VM16SH. Five speed gearbox. Clutch, wet multi-plate. Tyres, 2.50 x 19 front, 3 x 17 rear. Brakes, drum front and rear. Dimensions: Length, 1,860mm. Wheelbase, 1,210mm. Seat height, 765mm. Ground clearance, 225mm. Dry weight, 72kg. Fuel-tank capacity, 4.7 litres.

Yamaha FS1DX

Engine: Two stroke, single cylinder, reed valve, 49cc (40 x 39.7mm). Compression ratio, 6.6:1. Kick starter. Lubrication, Autolube. Carburettor, VM16SC. Four speed gearbox. Clutch, wet multi-plate. Tyres, 2.50 x 17 front and rear. Brakes, disc front, drum rear. Dimensions: Length, 1,790mm. Wheelbase, 1,160mm. Seat height, 770mm. Dry weight, 73kg. Fuel-tank capacity, 9 litres.

Yamaha PW50

Engine: Two stroke, single cylinder, 49cc (40 x 39.2mm). Compression ratio, 6:1. Max power, 2.7hp/5,500rpm. Max torque, 0.39kg/4,500rpm. CDI. Kick starter. Lubrication, Autolube. Carburettor, Mikuni VW12. Automatic gearbox, shaft drive. Front telescopic fork, rear pivoted fork. Tyres, 2.50 x 10 front and rear. Brakes, drum front and rear. Dimensions: Length, 1,245mm. Width, 575mm. Wheelbase, 855mm. Ground clearance, 105mm. Seat height, 485mm. Dry weight, 37kg. Fuel-tank capacity, 2 litres.

Yamaha V80

Engine: Two stroke, single cylinder, 79cc (47 x 45.6mm). Primary kick starter. Three speed semi-automatic gearbox. Clutch, wet multi-plate. Tyres, 2.25 x 17 front and rear. Brakes, drum front and rear. Dimensions: Length, 1,840mm. Wheelbase, 1,170mm. Seat height, 730mm. Dry weight, 78kg. Fuel-tank capacity, 5.3 litres.

Yamaha V50

Engine: Two stroke, single cylinder, reed valve, 49cc (40 x 39.7mm). Primary kick starter. Two speed automatic gearbox. Clutch, wet multi-plate. Tyres, 2.25 x 17 front and rear. Brakes, drum front and rear. Dimensions: Length, 1,840mm. Wheelbase, 1,170mm. Seat height, 730mm. Dry weight, 74kg. Fuel-tank capacity, 4.5 litres.

Yamaha Passola SA50

Engine: Two stroke, single cylinder, 49cc (40 x 39.2mm). Lubrication, Autolube. Kick starter. Two speed automatic gearbox. Tyres, 2.75 x 10 front and rear. Brakes, drum front and rear. Dimensions: Length, 1,580mm. Wheelbase, 1,115mm. Seat height, 715mm. Dry weight, 51kg. Fuel-tank capacity, 3 litres.

Yamaha AG100 Land Bike

Engine: Two stroke, single cylinder, 97cc (52 x 45.6mm). Compression ratio, 6.6:1. Max power, 8.5hp/6,250rpm. Max torque, 0.98kg/6,000rpm. CDI. Kick starter. Lubrication, Autolube. Five speed gearbox. Tyres, 2.75 x 19 front, 3.50 x 18 rear. Dimensions: Length, 2,110mm. Width, 930mm. Wheelbase, 1,315mm. Ground clearance, 235mm. Seat height, 800mm. Dry weight, 99kg. Fuel-tank capacity, 11 litres.

Yamaha YB100

Engine: Two stroke, single cylinder, rotary valve, 97cc (52 x 45.6mm). Compression ratio, 6.5:1. Max power, 10.1hp/8,000rpm. Lubrication, Autolube. Kick starter. Carburettor, VM20 SC x 1. Four speed gearbox. Clutch, wet multi-plate. Tyres, 2.50 x 18 front and rear. Brakes drum front and rear. Dimensions: Length, 1,915mm. Wheelbase, 1,180mm. Seat height, 786mm. Dry weight, 83.5kg. Fuel-tank capacity, 8.6 litres.

Yamaha YZ465

The production moto-crossers from Yamaha are extremely businesslike, powerful and, apart from the 125LC, air cooled. Most have monoshock rear suspension. Yamaha's is the most complete moto-cross range offered by the Japanese concerns. Two statistics of the YZ465, taken together, indicate the success of moto-cross technology: 52 bhp and 102kg dry weight . . .

Engine: Two stroke, single cylinder, 465cc (85 x 82mm). Compression ratio, 7.0:1. Max power, 52hp/7,000rpm. Max torque, 3.65kg-m/6,000rpm. Kick starter. Lubrication, 32:1 petrol/oil mixture. Six speed gearbox. Tyres, 3.00 x 21 front, 5.10 x 18 rear. Brakes, drum front and rear. Dimensions: Length, 2,175mm. Width 935mm. Height, 1,195mm. Wheelbase, 1,480mm. Ground clearance, 310mm. Seat height, 935mm. Dry weight, 102kg. Fuel-tank capacity, 9 litres.

Yamaha YZ250

Engine: Two stroke, single cylinder, 246cc (70 x 64mm). Compression ratio, 8.1:1. Max power, 40hp/8,000rpm. Max torque, 3.77kg-m/6,500rpm. Kick starter. Lubrication, 32:1 petrol/oil mixture. Six speed gearbox. Tyres, 3 x 21 front, 5.10 x 21 rear. Brakes, drum front and rear. Dimensions: Length, 2,155mm. Width, 935mm. Height, 1,195mm. Wheelbase, 1,455mm. Ground clearance, 310mm. Seat height, 935mm. Dry weight, 97kg. Fuel-tank capacity, 7.6 litres.

Yamaha DT250MX

Engine: Two stroke, single cylinder, seven port induction, 246cc (70 x 64mm). Compression ratio, 6.7:1. Max power, 23hp/6,000rpm. Kick starter. Lubrication, Autolube. Carburettor, VM28SS. Five speed gearbox. Clutch, wet multi-plate. Tyres, 3 x 21 front, 4 x 18 rear. Brakes, drum front and rear. Dimensions: Length, 2,145mm. Wheelbase, 1,415mm. Ground clearance, 245mm. Seat height, 860mm. Weight, 119kg. Fuel-tank capacity, 8 litres.

Yamaha YZ125LC

Engine: Two stroke, liquid cooled, single cylinder, 123cc (56 x 50mm). Compression ratio, 8.1:1. Max power 30hp/10,500rpm. Max torque, 2.07kg/10,250rpm. CDI. Kick starter. Lubrication, pre-mix (16:1). Six speed gearbox. Brakes, drum front and rear. Dimensions: Length, 2,140mm. Width, 880mm. Wheelbase, 1,450mm. Ground clearance, 345mm. Seat height, 945mm. Dry weight, 89kg. Fuel-tank capacity, 6.5 litres.

Yamaha YZ125

Engine; Two stroke, single cylinder, 123cc (56 x 50mm). Compression ratio, 8.3:1. Max power, 26.5hp/11,000rpm. Max torque, 1.80kg-m/9,000rpm. Kick starter. Lubrication, 32:1 petrol/oil mix-

ture. Six speed gearbox. Tyres, 3 x 21.4 front, 4 x 18 rear. Brakes, drum front and rear. Dimensions: Length, 2,115mm. Width, 950mm. Height, 1,215mm. Wheelbase, 1,430mm. Ground clearance, 340mm. Seat height, 940mm. Dry weight, 85kg. Fuel-tank capacity, 6.5 litres.

Yamaha YZ100

Engine: Two stroke, single cylinder, 98cc (50 x 50mm). Compression ratio, 8.4:1. Max power, 22hp/11,500rpm. Max torque, 1.40kg-m/9,500rpm. Kick starter. Lubrication, 32:1 petrol/oil mixture. Six speed gearbox. Tyres, 3 x 21 front, 4.10 x 18.4 rear. Brakes, drum front and rear. Dimensions: Length, 2,060mm. Width, 890mm. Height, 1,165mm. Wheelbase, 1,375mm. Ground clearance, 310mm. Seat height, 860mm. Dry weight, 84kg. Fuel-tank capacity, 5 litres.

Yamaha YZ80

Engine: Two stroke, single cylinder, 79cc (49 x 42mm). Compression ratio, 8.1:1. Max power, 17hp/11,500rpm. Max torque, 1.05kg-m/11,500rpm. Kick starter. Lubrication, 20:1 petrol/oil mix-

ture. Six speed gearbox. Tyres, 2.75 x 17.4 front, 3.60 x 14.4 rear. Brakes, drum front and rear. Dimensions: Length, 1,745mm. Width, 785mm. Height, 990mm. Wheelbase, 1,185mm. Ground clearance, 225mm. Seat height, 740mm Dry weight, 62kg. Fuel-tank capacity, 4.7 litres.

Yamaha YZ50

Schoolboy sport is big business these days, with plenty of organized competition for up-to-16-year-olds — and, by proxy, for their fathers who live out fantasies while footing the bill for these not-so-cheap mini-crossers. The YZ50 has a 'torque-induction' single-cylinder two-stroke engine and monoshock rear suspension. The bigger YZs, in 80 and 100cc form, are remarkably quick for their size.

Engine: Two stroke, single cylinder, 49cc (40 x 39.7mm). Compression ratio, 7.8:1. Max power 9hp/10,500rpm. Max torque, 0.62kg/10,000rpm. CDI. Kick starter. Lubrication, pre-mix. (16:1). Five speed gearbox. Tyres, 2.50 x 14 front, 3.00 x 12 rear. Brakes, drum front and rear. Dimensions: Length, 1,510mm. Width, 715mm. Wheelbase, 1,025mm. Ground clearance, 195mm. Seat height, 630mm. Dry weight, 50kg. Fuel-tank capacity, 3 litres.

Yamaha IT465

The IT series are enduro bikes. They have headlamps, so you might wonder how they differ from the DT bikes. Principally in having more power . . . in being more 'professional'. Look on them as slightly detuned moto-crossers.

Engine: Two stroke, single cylinder, 465cc (85 x 82mm). Compression ratio, 7.1:1. Max power, 46hp/6,500rpm. Max torque, 5.3kg/5,500rpm. CDI. Kick starter. Lubrication, pre-mix. Carburettor, Mikuni VM38. Five speed gearbox. Front telescopic forks, monocross rear. Tyres, 3 x 21 front, 5.60 x 17 rear. Brakes, drum front and rear. Dimensions: Length, 2,205mm. Width, 890mm. Wheelbase, 1,475mm. Ground clearance, 295mm. Seat height, 925mm. Dry weight, 120kg. Fuel-tank capacity, 13 litres.

Yamaha IT250

Engine: Two stroke, single cylinder, 246cc (70 x 64mm). Compression ratio, 7.9:1. Max power, 37hp/7,000rpm. Max torque, 3.7kg/7,000rpm. CDI. Kick starter. Lubrication, pre-mix. Carburettor, Mikuni VM36SS. Six speed gearbox. Front telescopic forks, rear monocross. Tyres, 3 x 21 front, 5.10 x 18 rear. Brakes, drum front and rear. Dimensions: Length, 2,205mm. Width, 890mm. Wheelbase, 1,450mm. Ground clearance, 295mm. Seat height, 925mm. Dry weight, 106kg. Fuel-tank capacity, 13 litres.

Yamaha TZ500

The TZ series of racers are sold ready for the track. Well . . . let's say then that you would need some-thing of the knowhow of a Kenny Roberts to improve them significantly. The 500 produces more power than ever, at 110 bhp, and the chassis has been modified. Exhaust-timing variation is provided by the YPVS (Yamaha Power Valve System). New for 1981: special cutaway throttle valve for the carburettors; improved exhaust-pipe setup; more sensitive damping for the shock absorber controlling the monocross rear suspension; increase in front-wheel rim size.

Engine: Two stroke, four cylinder YPVS, 497cc (56 x 50mm). Compression ratio, 7.9:1. Max power, 110hp/10,500rpm. Max torque, 7.9kg/10,250rpm. Push start. Lubrication, 15:1 petrol/oil mixture. Six speed gearbox. Tyres, 3.25 x 18 front, 4.00/5.75 x 18 rear. Brakes, disc front and rear. Dimensions: Length, 2,020mm. Width, 500mm. Height, 965mm. Wheelbase, 1,365mm. Ground clearance, 120mm. Seat height, 900mm. Net weight, 139kg. Fuel-tank capacity, 31.5 litres.

Yamaha TZ750

Engine: Two stroke, four cylinder, 747cc (66.4 x 54mm). Compression ratio, 7.3:1. Max power, 100hp/10,000rpm. Max torque, 7.2kg/9,500rpm. Push start. Lubrication, 15:1 petrol/oil mixture. Six speed gearbox. Tyres, 3.25 x 18 front, 3.50/5.25 x 18 rear. Brakes, disc front and rear. Dimensions: Length, 2,015mm. Width, 637mm. Height, 935mm. Wheelbase, 1,390mm. Ground clearance, 127mm. Seat height, 730mm. Net weight, 155kg. Fuel-tank capacity, 24 litres.

Yamaha TZ250

Several changes to the very successful 250 racers were brought in for 1981: lighter engine (with YPVS), enlarged transfer ports, TZ500-type carburation, separate cylinder and cylinder head, new steering geometry . . .

Engine: Two stroke, twin cylinder, 247cc (54 x 54mm). Compression ratio, 7.8:1. Max power, 46hp/10,000rpm. Max torque 3.29kg/10,000rpm. Push start. Lubrication, 15:1 petrol/oil mixture. Six speed gearbox. Tyres, 3.00 x 18 front, 3.50 x 18 rear. Brakes, disc front and rear. Dimensions: Length, 1,935mm. Width, 630mm. Height, 950mm. Wheelbase, 1,320mm. Ground clearance, 170mm. Seat height, 730mm. Dry weight, 106.5kg. Fuel-tank capacity, 23.5 litres.

Yamaha TZ125

Engine: Two stroke two cylinder, 123cc (56 x 50mm). Compression ratio, 7.9:1. Max power, 30hp/12,000rpm. Max torque, 1.85kg/11,500rpm. Push start. Lubrication, 15:1 petrol/oil mixture. Six speed gearbox. Tyres, 2.50 x 18 front and rear. Brakes, disc front and rear. Dimensions: Length, 1,790mm. Width, 520mm. Height, 895mm. Wheelbase, 1,205mm. Ground clearance, 155mm. Seat height, 685mm. Dry weight, 72kg. Fuel-tank capacity, 9.5 litres.

ACKNOWLEDGEMENTS are due to the following for providing, variously, copy, source material and photographs:
J.W.E. Kelly
B.R. Nicholls
Paul Boland
David Dixon
M.R. Wigan
Maurice Spalding Publicity
H.N. Jeffery
Peter Richardson
Kawasaki Information Service
Chris Bryant
Mitsui
Yamaha Motor
Phil Manzano
Brian Fowler
Jeff Clew
V.H. Willoughby
Alan Davidson
Suzuki Information Service